Surveys and Tutorials in the Applied Mathematical Sciences

Volume 18

Featuring short books of approximately 80-200pp, Surveys and Tutorials in the Applied Mathematical Sciences (STAMS) focuses on emerging topics, with an emphasis on emerging mathematical and computational techniques that are proving relevant in the physical, biological sciences and social sciences. STAMS also includes expository texts describing innovative applications or recent developments in more classical mathematical and computational methods.

This series is aimed at graduate students and researchers across the mathematical sciences. Contributions are intended to be accessible to a broad audience, featuring clear exposition, a lively tutorial style, and pointers to the literature for further study. In some cases a volume can serve as a preliminary version of a fuller and more comprehensive book.

Tomoyuki Miyaji · Shin-Ichiro Ei ·
Masayasu Mimura

A Billiard Problem in Nonlinear Dissipative Systems

Springer

Tomoyuki Miyaji
Department of Mathematics
Kyoto University
Kyoto, Japan

Masayasu Mimura
Hiroshima University
Hiroshima, Japan

Meiji University
Tokyo, Japan

Shin-Ichiro Ei
Department of Mathematics
Josai University
Sakado, Japan

Hokkaido University
Sapporo, Japan

ISSN 2199-4765 ISSN 2199-4773 (electronic)
Surveys and Tutorials in the Applied Mathematical Sciences
ISBN 978-981-95-7560-2 ISBN 978-981-95-7561-9 (eBook)
https://doi.org/10.1007/978-981-95-7561-9

Mathematics Subject Classification: 37N99, 35B32, 35Q92, 35K57, 37G10

This Springer imprint is published by the registered company Springer Nature Singapore Pte Ltd.
The registered company address is: 152 Beach Road, #21-01/04 Gateway East, Singapore 189721, Singapore

Preface

The classical billiard problem, which assumes a conservative system of repeated elastic collisions of a moving ball or balls within a bounded domain, has been extensively studied by the mathematical community. This book models the billiard-like motion of a self-propelling disk floating on the surface of a rectangular pool. While this book's topic concerns billiard modeling, it departs from the classical paradigm of a conservative system, because the system we model is, in fact, a dissipative one. The self-propelling properties of a camphor disk floating on water are well known in nonlinear science. Laboratory experiments conducted by a former student of the third author demonstrate that, unlike classical billiards, a camphor disk has the following properties:

(i) It reflects even though it does not collide with the boundary.
(ii) The angle of reflection is greater than that of incidence, quite unlike elastic reflection.

As such, a camphor disk's behavior differs from that of classical billiards. This book aims to show how such a self-propelling disk behaves in general.

We propose three types of mathematical model of the camphor disk: a moving-boundary (MB) model, a particle model, and a discrete-time model. To begin with, the MB model offers a mathematical account of the mechanism that drives camphor motion. Then, to show that the MB model satisfies properties (i) and (ii) above, we derive the particle model from the MB model. The particle model describes the position and velocity of the disk under the assumption that the motion is sufficiently slow. Classical billiards exhibit simple behavior in a rectangular domain, whereas the particle model exhibits additional, complex behavior depending on the aspect ratio of the rectangle. To understand such complex behavior, we derive a discrete-time model describing the motion of a particle in terms of straight lines connecting sites of reflection, at which the direction of motion changes according to an asymmetric reflection rule. Applying dynamical systems theory, bifurcation theory, and numerical simulations, we find that a self-propelling disk exhibits a variety of complex motions due to the asymmetry of incident and reflected angles, even if the domain is a rectangle. If we were to think of this system as an actual game of billiards, we would observe that the future path of a billiard ball in a nonlinear dissipative system would mainly be determined by factors and perturbations which would have little to do with the player's experience. In contrast, player's experience more naturally determines the paths balls take in a game of classical billiards.

This book is organized as follows. Chapter 1 is devoted to a general introduction to the problems under consideration and a brief discussion of experimental arrangements, but without going into great detail. The three mathematical models are then discussed in Chaps. 3, 4, and 5. Before entering into the details of each model, in Chap. 2, we review some basic knowledge from the theory of dynamical systems required in the following chapters. In Chap. 3, the MB model is introduced. This model is constructed on the basis of knowledge gained from experimental observation. In addition, the center manifold theory for the MB model is developed, which enables us to derive the particle model as a bifurcation equation under the assumption that the system parameter is sufficiently close to the value at which the standing spot solution of the MB model loses stability. We also consider the interactions of two disks. In Chap. 4, the particle model is studied for the entire plane, half plane, and rectangular domains. In Chap. 5, the discrete-time model is derived and studied to understand the outcome of our dissipative billiard in a bounded domain. Finally, concluding remarks are given in Chap. 6.

The content of this book is based on the bachelor's and master's theses of the third author's supervised students and research conducted by the present authors. Unfortunately, these theses have not been published and are inaccessible. In compiling the book, we have reorganized the theory rather than directly reproducing what has been previously written. Nevertheless, we cannot overemphasize that the students made essential contributions to this study.

As the title of this book suggests, "a billiard problem" on a rectangular domain is studied in this book. In classical billiard problems, the specular reflection rule is fixed, and the geometry of the billiard table plays an essential role. It is, of course, an interesting problem to study nonspecular reflections on billiard tables with different shapes. Nevertheless, we concentrate on the case of a rectangular domain. In our case, the reflection rule is more complex, and diverse behaviors are observed even if the domain is a rectangle, and the results obtained for a rectangular domain will be instructive for general domains. We present a few examples of domains other than rectangles in Chaps. 4 and 5.

On a somber note, it is truly regrettable that Masayasu Mimura, one of the authors of this book, passed away on April 8, 2021. The first and second authors express deep sadness and profound regret at losing him. The original idea for this book arose from his detailed observations of the motion of camphor disks, both in actual experiments and in numerical simulations of mathematical models, and in particular, his discovery of the property (ii), that the angle of reflection is greater than the angle of incidence for the camphor disk. This property (ii) is quite different from the usual billiard model assumption; at first, the surviving authors did not understand what it meant. One day, Prof. Mimura called upon us to help explain the existence of stable periodic solutions in a square domain. It was his unique way of understanding this phenomenon, which we felt that no one else but he could have noticed. That was about twenty years ago. Since then, there have been many studies of dissipative billiards in rectangles of various aspect ratios. It has become clear that many properties of dissipative systems, such as the emergence of chaos, are exemplified by these systems. Based on these studies, Prof. Mimura took the lead in the writing of this book. He was striving to develop a mathematical understanding of the essential structures of these phenomena through modeling. In that sense, this book reflects his ideas and way of thinking. While writing this book, he had already thought about and prepared many things, including the introduction and overall structure, but he passed away in the middle of this work. Therefore, we have attempted to reflect his wishes and develop his line of thought as much as possible in the writing of this book. However, those who knew him well may feel that this work is rather different from what he originally intended. We admit that we are not good enough. In closing, we would like to express our deepest gratitude to Prof. Mimura for guiding us through various discussions and joint work. In particular, Prof. Mimura was the supervisor of the second author. Thus, we express our profound sense of loss once again.

April 2025 The Authors

Acknowledgments

The authors would like to thank Prof. Shunsuke Izumi of Hiroshima University for his extensive advice and support for the students' experiments on camphor disks, which are described in Chap. 1. Professor Izumi showed the third author, Masayasu Mimura, and Mimura's former students, who had no experience with laboratory experiments, how to construct and use the experimental apparatus. The third author received this assistance in his role as an academic staff member at the Graduate School of Mathematical and Life Science at Hiroshima University, a school that promoted interdisciplinary collaboration among researchers in biology, mathematics, and chemistry.

The first author was supported by JSPS Kakenhi 16K17649 and 19K03626, and the second author was supported by JSPS Kakenhi 22H01129 and JST CREST JPMJCR14D3.

The authors thank Enago (www.enago.jp) for the English language review.

The authors wish to thank the anonymous reviewers for their valuable comments which improved the quality of the manuscript. We would like to thank Mr. Masayuki Nakamura of Springer Japan for his continuous support.

Finally, we express our special gratitude to Dr. Robert Sinclair for his careful reading of the manuscript, his linguistic correction, and very helpful comments.

Contents

Chapter 1
Introduction

Abstract Inspired by the self-propelled, billiard-like motion of a camphor disk floating on the surface of water, we motivate and introduce three mathematical models. Our camphor disk moves in a straight line and reflects as it approaches the boundary of the tank, somewhat like a billiard ball, but exhibits nonspecular reflection. It loses speed on reflection, but then it recovers its speed. How does it behave if it is confined to a square domain? In contrast to classical billiards, it seems that its asymptotic orbit is determined almost independently of the initial condition in the case of a camphor disk.

1.1 Introduction

The spontaneous formation of spatiotemporal patterns is one of the most fascinating phenomena observed in the behavior of nonlinear dissipative systems. It is observed in many areas of the natural and social sciences at various spatiotemporal scales; for instance, the Belousov-Zhabotinsky reaction, the development of microbial colonies, spatiotemporal distributions of species, invasions, and waves of urban traffic congestion and decongestion, among many more such phenomena. Although a macroscopic pattern can result from the collective behavior of individuals, apparently spontaneous motions of each individual may also be generated by self-organization. When a scientist faces such phenomena, mathematical modeling often proves indispensable. There can be many approaches to mathematical modeling, depending on how the mathematical scientist would like to understand the phenomenon under investigation. For example, suppose spatial patterns are formed by a very large number of individuals. In that case, a macroscopic model based upon partial differential equations, such as reaction-diffusion systems, can be helpful. Such models describe the evolution of the spatial distribution of a population continuum over time, rather than the dynamics of each individual. One may model population continua from the standpoints of partial differential equations, dynamical systems theory, and numerical analysis. In some cases, however, continuum approximations may not be suitable. In such cases,

T. Miyaji et al., *A Billiard Problem in Nonlinear Dissipative Systems*,
Surveys and Tutorials in the Applied Mathematical Sciences 18,
https://doi.org/10.1007/978-981-95-7561-9_1

one may be able to describe the dynamics of each individual in a large population. Such a model is called a microscopic or individual-based model. A microscopic model describes the motions of individuals and the interactions among them. In a microscopic model, each individual is usually regarded as a particle that may move spontaneously. Such a particle is called a self-propelling, or self-propelled, particle.

Active matter is a more general category, which includes self-propelling particles as a subset. The study of active matter is one of the most rapidly growing research fields in nonlinear physics. In [11], the following definition is supplied: "Active matter is commonly defined as soft systems with built-in dispersed and local conversion of energy into forces and motions." Vicsek et al. [83] have succeeded in characterizing coherent collective motion as a phase transition in a simple agent-based model. Their pioneering work stimulated the development of active matter research, and the field rapidly broadened to include the collective motion of fish, birds, insects, and even nonliving objects [84]. The term nonliving is preferred here to the more usual term inanimate, as in the cases under investigation, the nonliving objects are self-propelled and thus animate, though nonliving. In the self-propelled motion of an individual, the mode of motion and shape deformation of the individual are closely related to each other, especially for swimming bacteria and oil droplets floating on aqueous solutions [59]. There are two major questions that need to be addressed:

1. What kinds of collective motion are possible for given individuals which exhibit certain motions and interactions?
2. What kinds of motions and interactions are possible for each system?

Studies designed to address these questions have been rapidly growing in number.

Let us now consider the motion of a camphor scrapping upon the surface of a body of water. This is an example of a simple nonliving object that nonetheless exhibits self-propelled motion. Camphor ($C_{10}H_{16}O$) is a chemical material which can take the form of a white crystalline powder [60]. It has a characteristic odor and is commonly used as an ingredient of aromatics, pesticides or moth-proofing. It sublimates at room temperature and its solubility in water is low. If a scrapping of camphor is placed on the surface of a body of water, it spontaneously begins to move as though it were dancing [76].

The mechanism of camphor motion is explained as follows [76]: the surface tension decreases due to camphor molecules spreading on the water surface. This invariably results in the surface tension around the camphor scrapping becoming inhomogeneous. As a result, the camphor scrapping is pulled toward the adjacent area where the surface tension is highest. Since camphor sublimates at room temperature, camphor molecules are continuously removed from the water surface. As the water surface becomes free of camphor molecules, more camphor molecules can be supplied from the scrapping to the water surface. Thus, the camphor scrapping is spontaneously self-propelled and can keep moving until it evaporates. If the camphor scrapping has a complex shape, then the surface tension distribution will also be complicated, and its motion will be essentially unpredictable. But what happens if the floating camphor particle is not just a random and asymmetrical scrapping, but a disk?

1.2 Motion of a Camphor Disk

From an elementary mathematical or geometrical viewpoint, if a camphor particle is radially symmetric and uniform, then its symmetry should prelude self-propulsion due to the surface tension effect. On the other hand, from a practical viewpoint, one may expect that the camphor disk will become self-propelling, on the grounds that it is impossible to create such a disk and to put it on a water surface without creating a slight asymmetry in practice. Which opinion is correct?

As this question of the motion of camphor disk seemed elementary, it has attracted little attention. Nonetheless, a former undergraduate student of one of the present authors in the school of mathematics at Hiroshima University decided to investigate the motions of camphor disks on water by experimental means [32]. The student's experimental facility is shown in Fig. 1.1. In the student's experiment, borneol was used instead of camphor. Borneol ($C_{10}H_{18}O$) and camphor are closely related in chemical terms. Borneol, which is also called Borneo camphor, has a camphor-like odor, sublimates at room temperature and is insoluble in water [61], and it exhibits similar self-propelling behavior when floated on water. A square tank, 20 cm on a side, was filled with water to 7 cm depth, and a borneol disk of 1.35 cm diameter and 0.05 cm thickness was floated on the water surface. The motion of the borneol disk was monitored by a video camera. The disk was made of borneol by applying 600 kg/cm^2 pressure to borneol with a micro tablet molding machine, which also generated a smooth surface. The water on which the borneol was floated was both distilled and deionized.

The result was as follows. When the purity of the borneol was low, the disk soon stopped moving. On the other hand, when the purity was sufficiently high, it continued to move. Each result fulfilled one of the a priori predictions with the difference depending, in practice, on the purity of the borneol.

In addition, some interesting phenomena of a less obvious nature were observed. Figure 1.2 presents a trajectory of the borneol disk used in the experiment. The

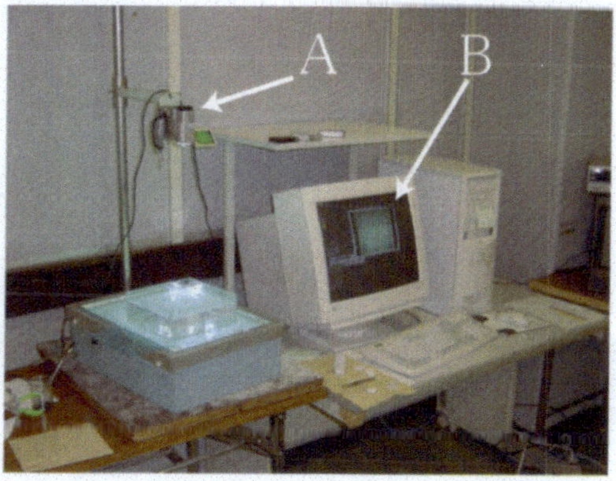

Fig. 1.1 The experimental facility used in [32]. The experiment was recorded by the video camera at A and monitored by the display at B to produce a movie

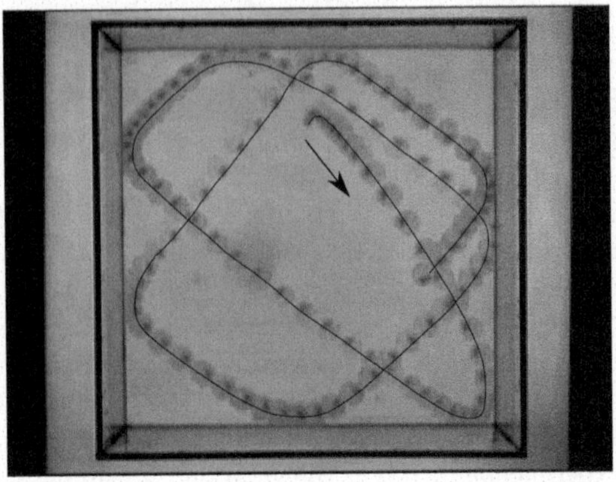

Fig. 1.2 A trajectory of a camphor disk of an experiment by [32]

borneol disk moved almost in a straight line, it then changed its direction by a certain angle when it got close to the tank wall. It then unavoidably approached the tank wall at another location, and was reflected again. At first glance, this motion is similar to that of a billiard ball.

On the other hand, the reflection of the borneol disk was not specular. That is, the angle of reflection was different from that of incidence. Moreover, the disk changed direction without touching the tank wall, a fact which means that the reflection of the borneol disk was not caused by rigid body collision but by some other mechanism. These observed behaviors of the borneol disk are quite dissimilar to those of the classical mathematical billiard problem. It is expected that the trajectory of a borneol disk is quite different from that of the classical mathematical billiard problem which assumes boundary contact and elastic reflection.

Unfortunately, the student's experiment has not been independently reproduced. Nevertheless, we proceed to our mathematical study inspired by the students's experiment. As we will see below, a mathematical model suggests that the mechanism that drives self-propelled motion of a camphor disk can lead to billiard-like motion and collision-free reflection. The authors will be happy if these experiments are repeated by chemists in the future.

1.3 Moving Boundary Model

Let us consider a mathematical model that describes camphor motions. We assume that the shape of the camphor disk never changes. In reality, since camphor molecules spread out onto the water, the shape of camphor disk may change over time. However the shape does not change very much even over 1–2 h, which is a much longer time than the time required for a useful experiment.

Let $r_0 > 0$ be arbitrarily fixed, and let $B(\boldsymbol{p}(t), r_0)$ be a disk of radius $r_0 > 0$ centered at $\boldsymbol{p}(t) \in \mathbb{R}^2$, where $\boldsymbol{p}(t)$ is a function of time $t \in \mathbb{R}$. The disk $B(\boldsymbol{p}(t), r_0)$

represents a camphor disk with radius r_0 located at $\boldsymbol{p}(t)$. Let ρ and h_0 be the density and the thickness of the camphor disk, respectively. The mass of the camphor disk is given by $m = \rho\pi r_0^2 h_0$. Let $c(t, \boldsymbol{x})(\boldsymbol{x} = (x, y) \in \mathbb{R}^2)$ be the concentration of camphor molecules on the water surface. The motion of the camphor disk is caused by surface tension, which is a function of $c(t, \boldsymbol{x})$. In addition, the disk experiences resistance caused by the viscosity of the water. This resistance is proportional to the contact area πr_0^2 and the velocity $\dot{\boldsymbol{p}}(t)$. From these assumptions, as we will explain in Chap. 3, the center position $\boldsymbol{p}(t)$ of the disk is governed by the following equation:

$$m\ddot{\boldsymbol{p}}(t) = \int_{\partial B(\boldsymbol{p}(t), r_0)} \Gamma(c(t, \boldsymbol{x}))\boldsymbol{n}ds - \mu\pi r_0^2 \dot{\boldsymbol{p}}(t), \tag{1.1}$$

where $\boldsymbol{n}$ is the outward unit normal vector on $\partial B(\boldsymbol{p}(t), r_0)$, $\Gamma(c)$ is a monotone decreasing function that represents the surface tension, and μ is the viscosity coefficient. See Fig. 1.3. On the other hand, the concentration $c(t, \boldsymbol{x})$ of the camphor molecules on the water surface is given by

$$c_t = D\Delta c - \alpha c \chi_{\mathbb{R}^2 \setminus B(\boldsymbol{p}(t), r_0)} + F(c)\chi_{B(\boldsymbol{p}(t), r_0)}, \tag{1.2}$$

where $\alpha > 0$ and $D > 0$ are constants, $\Delta = \partial_x^2 + \partial_y^2$ is the Laplacian, and $\chi_{B(\boldsymbol{p}(t), r_0)}$ and $\chi_{\mathbb{R}^2 \setminus B(\boldsymbol{p}(t), r_0)}$ are defined by

$$\chi_{B(\boldsymbol{p}(t), r_0)} = \begin{cases} 1 & (x \in B(\boldsymbol{p}(t), r_0)), \\ 0 & (x \notin B(\boldsymbol{p}(t), r_0)), \end{cases} \quad \chi_{\mathbb{R}^2 \setminus B(\boldsymbol{p}(t), r_0)} = \begin{cases} 0 & (x \in B(\boldsymbol{p}(t), r_0)), \\ 1 & (x \notin B(\boldsymbol{p}(t), r_0)), \end{cases}$$

respectively. The right hand side of (1.2) describes the diffusion, sublimation and supply of camphor molecules (see Fig. 1.4). The function $F(c)$ is a nonnegative monotone nondecreasing function. Here, we assume that $F(c)$ is a constant function $F(c) \equiv a_0 > 0$ for simplicity. The constant a_0 corresponds to the purity of the camphor disk.

Assuming that the camphor disk is very light, we can replace (1.1) by

$$\dot{\boldsymbol{p}}(t) = \frac{1}{\mu\pi r_0^2} \int_{\partial B(\boldsymbol{p}(t), r_0)} \Gamma(c(t, \boldsymbol{x}))\boldsymbol{n}ds. \tag{1.3}$$

Since the boundary of the moving disk $B(\boldsymbol{p}(t), r_0)$ moves as time passes, we call (1.2)–(1.3) a moving boundary (MB) model.

Let us investigate the MB model by numerical simulation. We will use the following parameter values and function $\Gamma(c)$ unless otherwise stated:

$$D = 0.2, \quad \mu = 0.35, \quad \alpha = 1.0,$$

$$r_0 = 1.0, \quad a_0 = 1.0, \quad \Gamma(c) = \frac{1}{1 + 2c}.$$

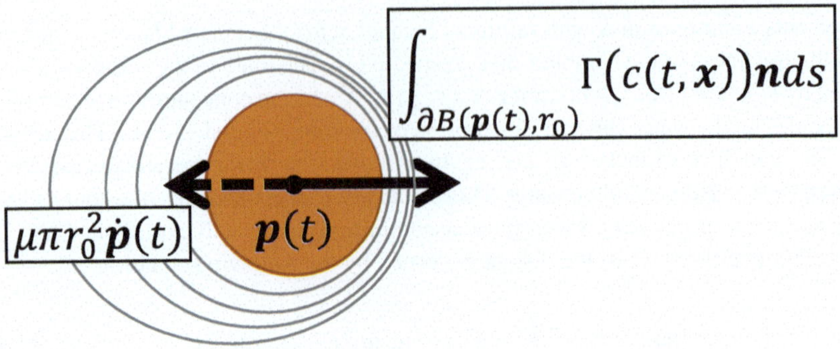

Fig. 1.3 Illustration of (1.1). The camphor disk tends to move in the direction of higher surface tension, while it is subject to viscous resistance. The gray curves represent level sets of $c(t, \boldsymbol{x})$

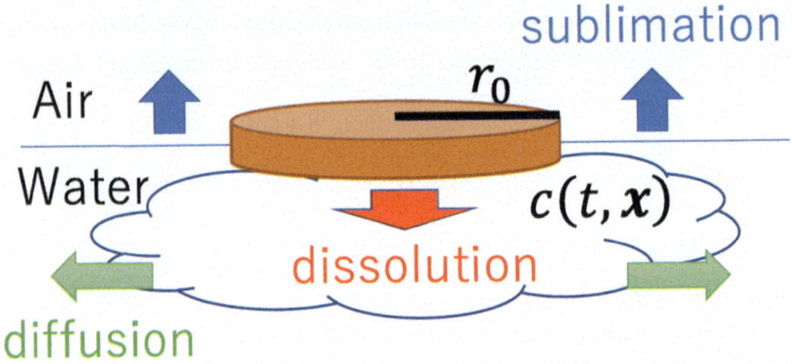

Fig. 1.4 Illustration of (1.2). Camphor molecules are supplied from the camphor disk to the surface of water and diffuse. They are removed from the surface of water by sublimation

These parameter values are not fitted to the above experiments. Here, we just observe qualitative properties of a typical solution of the MB model. Our method of numerical simulation is described in Sect. 3.2.

By numerical simulation of the MB model, we can observe that the disk moves and tends to uniform straight line motion. Figure 1.5 represents a disk moving from left to right along with the contour lines of $c(t, \boldsymbol{x})$, and Fig. 1.6 is a slice at $y = 0$. The center of the disk is located to the right of the extremum point of $c(t, \boldsymbol{x})$. As the camphor concentration in front of the disk (to the right) is lower than that behind it, the disk moves to the right. In addition, the disk keeps this positional relationship with the contour lines. That is, uniform motion of the disk is characterized by a traveling spot solution of the moving boundary model. To describe this situation mathematically, we may assume that there exists a profile function $S : \mathbb{R}^2 \to \mathbb{R}$ and a vector $\boldsymbol{v} \in \mathbb{R}^2$ such that

$$c(t, \boldsymbol{x}) = S(\boldsymbol{x} - t\boldsymbol{v}), \quad \boldsymbol{p}(t) = t\boldsymbol{v}.$$

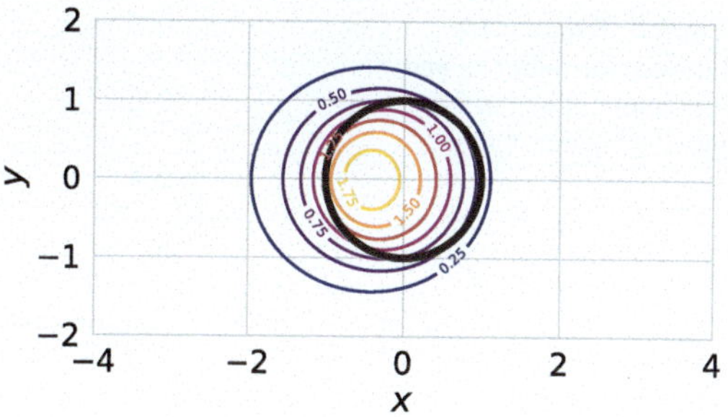

Fig. 1.5 A disk moving from left to right (black circle) and the contour lines of $c(t, \boldsymbol{x})$ (colored curves)

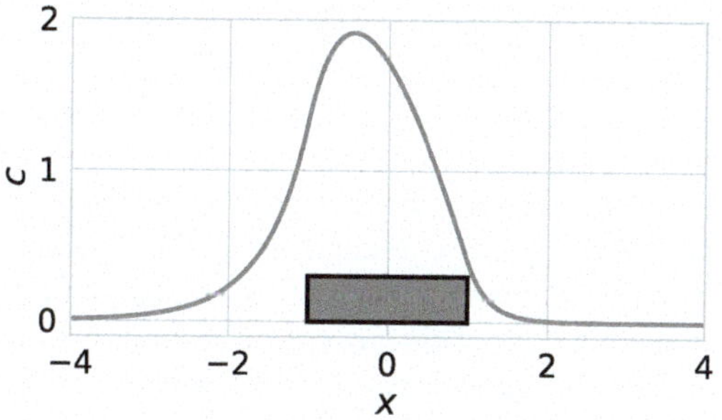

Fig. 1.6 A slice of Fig. 1.5 at $y = 0$. The gray curve is $c(t, x)$, and the rectangle corresponds to the disk

Furthermore, we can observe the dependence of the disk's motion upon the purity of camphor from which the disk is made. Figure 1.7 presents the relationship between purity and asymptotic speed. Here, the asymptotic speed was calculated as the value of $\dot{\boldsymbol{p}}(t)$ by solving the initial-boundary value problem of the MB model. At this parameter setting, there is a threshold at $a_0 \approx 0.24$. The speed tends to zero if $a_0 < 0.24$, while it tends to some positive constant if $a_0 > 0.24$. Thus, the model suggests that the disk can keep moving only if the camphor purity is above the threshold. It should be noted that there is another threshold value. It seems that the asymptotic speed becomes zero again for $a_0 > 1.21$. We have chosen the parameter values above so that the asymptotic speed becomes positive. If one chooses different values of D, μ, etc., then the threshold value of a_0 can be different. See Chap. 3 for details.

One may wish to control the disk's radius r_0 as a parameter. At first we expected that the speed was monotone increasing with respect to r_0. As shown in Fig. 1.8,

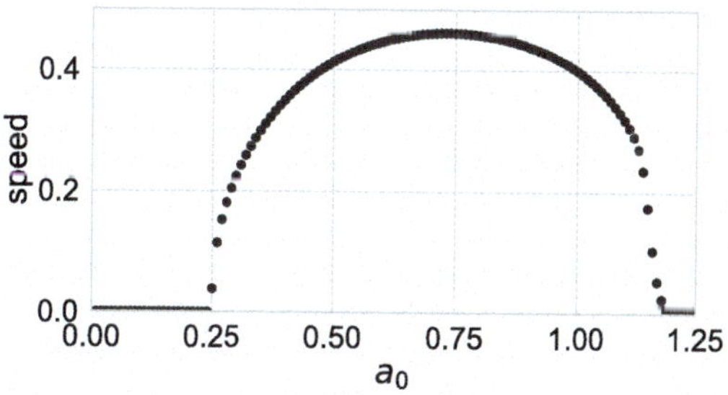

Fig. 1.7 Relationship between the purity a_0 and asymptotic speed in the moving boundary model

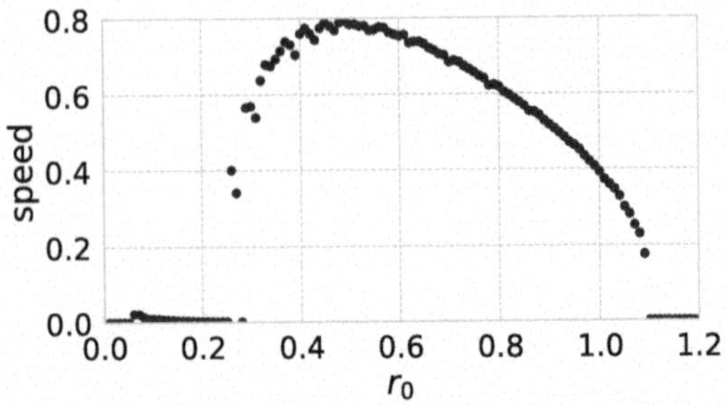

Fig. 1.8 Relationship between the radius r_0 and asymptotic speed in the moving boundary model

the asymptotic speed seems to depend on r_0 in a more complicated manner. This observation can be explained as follows. Note that the factor $1/(\pi r_0^2)$ appears on the right hand side of (1.3). A smaller disk is subjected to a weaker resistance from the viscosity of the water. If the disk is too small, then it cannot supply camphor molecules sufficient for self-propulsion. On the other hand, a larger disk needs more camphor supply to overcome resistance. Therefore, there is an interval of r_0 in which the speed is monotone decreasing. It should be noted that r_0 also affects the contour integral in (1.4). This prevents us from providing a simple interpretation.

Next, let us observe the relationship between angles of incidence and reflection. Figure 1.9 presents a trajectory of $\boldsymbol{p}(t)$ on a rectangular domain $[-80, 80] \times [-20, 20]$. Here, the homogeneous Neumann boundary condition is imposed for $c(t, \boldsymbol{x})$. The trajectory is vertical at first, but it gradually tends towards being the horizontal. This suggests that the angle of reflection is larger than that of incidence. This is an important property, which has also been observed experimentally.

Figure 1.10 shows the time dependence of the speed $|\dot{\boldsymbol{p}}(t)|$ for the same simulation as Fig. 1.9. The speed tends to some constant when the disk is near the middle of the domain. Although the disk decelerates when a reflection occurs, it accelerates and recovers the constant speed if it is away from the boundary.

Figure 1.11 shows trajectories of $\boldsymbol{p}(t)$ on $[-80, 80] \times [-20, 20]$ for $a_0 = 0.3, 0.4$, and 1.0. In each figure, the same initial condition was imposed. Figure 1.12 shows the time dependence of $|\dot{\boldsymbol{p}}(t)|$ for the simulation of Fig. 1.11. For $a_0 = 0.3, 0.4$, and

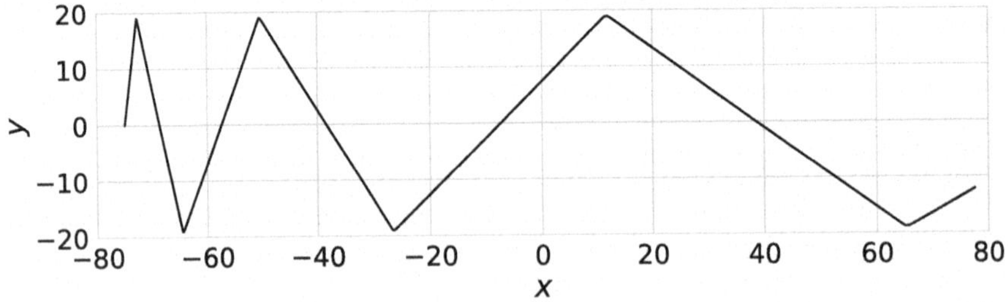

Fig. 1.9 A trajectory of $\boldsymbol{p}(t)$ of the moving boundary model on a rectangular domain $[-80, 80] \times [-20, 20]$. Figure adapted from [15, Fig. 3], used with permission of the American Institute of Mathematical Sciences, from Ei et al. [15]; permission conveyed through Copyright Clearance Center, Inc.

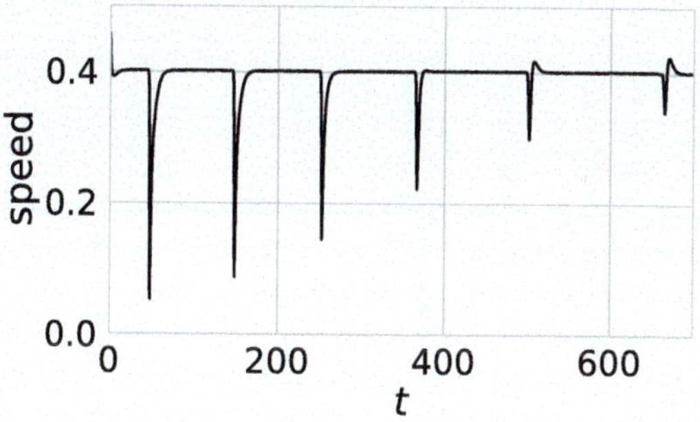

Fig. 1.10 Variation with time of $|\dot{\boldsymbol{p}}(t)|$ in the simulation of Fig. 1.9

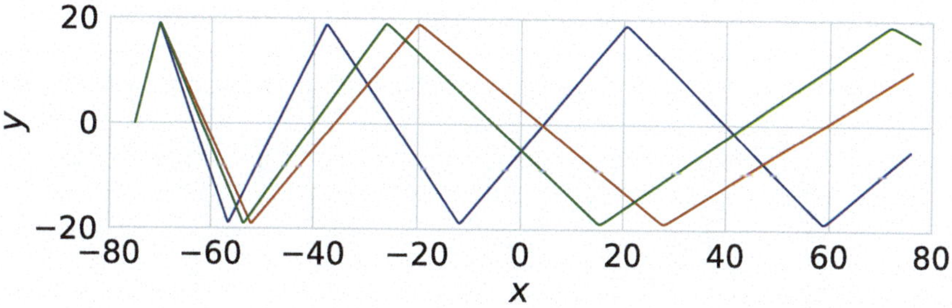

(a)

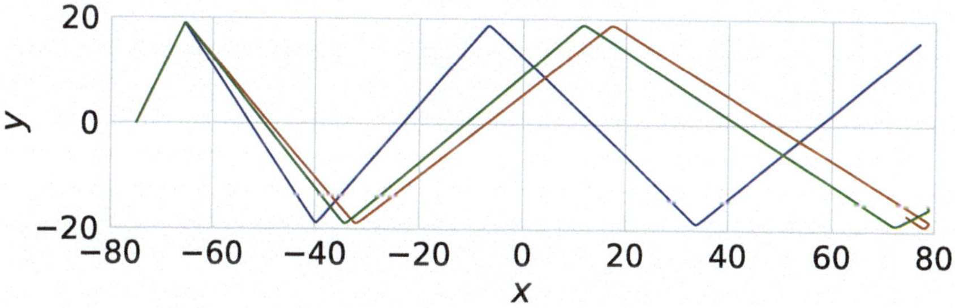

(b)

Fig. 1.11 Trajectories of $\boldsymbol{p}(t)$ of the MB model on a rectangular domain $[-80, 80] \times [-20, 20]$. In each figure, the trajectories with the same initial condition but with different a_0 are plotted. The blue, red, and green lines correspond to $a_0 = 0.3$, 0.4 and 1.0, respectively. Figure adapted from [15, Fig. 3], used with permission of the American Institute of Mathematical Sciences, from Ei et al. [15], permission conveyed through Copyright Clearance Center, Inc.

1.0, the asymptotic speed was monotone increasing in this order. In all cases, the angle of reflection is larger than that of incidence. We also calculated trajectories for $a_0 = 0.8$ and 0.9, but the disk hit the boundary in those cases.

As the experiment shown in Fig. 1.2, the disk does not collide with the boundary in the simulation of Fig. 1.9. (It should be remarked that the disk may still collide

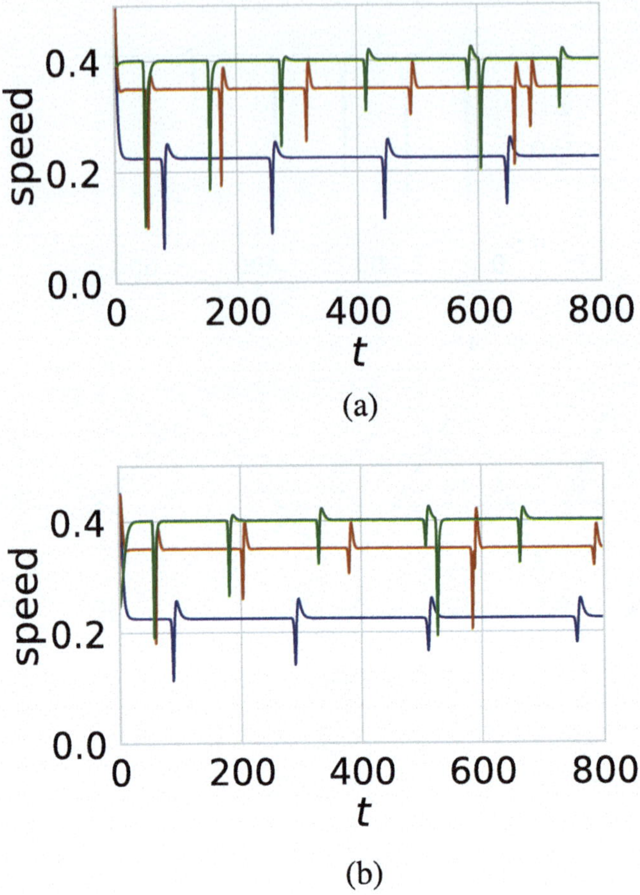

Fig. 1.12 Time dependence of $|\dot{\boldsymbol{p}}(t)|$ for the simulations of Fig. 1.11. The blue, red, and green lines correspond to $a_0 = 0.3$, 0.4 and 1.0, respectively

with the boundary if it is moving too fast to be turned back by surface tension forces. This will be briefly mentioned in Chap. 3. Our analytical focus here is on motion that is sufficiently slow to ensure that no boundary collision occurs.) Let us now observe a reflection near the boundary. See Fig. 1.13, which provides snapshots of the disk and contour of $c(t, \boldsymbol{x})$. As the disk approaches the boundary, the concentration of camphor molecules becomes relatively high in the area between the disk and boundary. This ensures that the surface tension behind the disk becomes relatively high. Therefore, the disk tends to move away from the boundary.

To summarize, we have simulated the motion of a camphor disk using the moving boundary model, and have confirmed the following properties:

1. The disk moves in straight lines between reflections.
2. The disk reflects without colliding with the boundary.

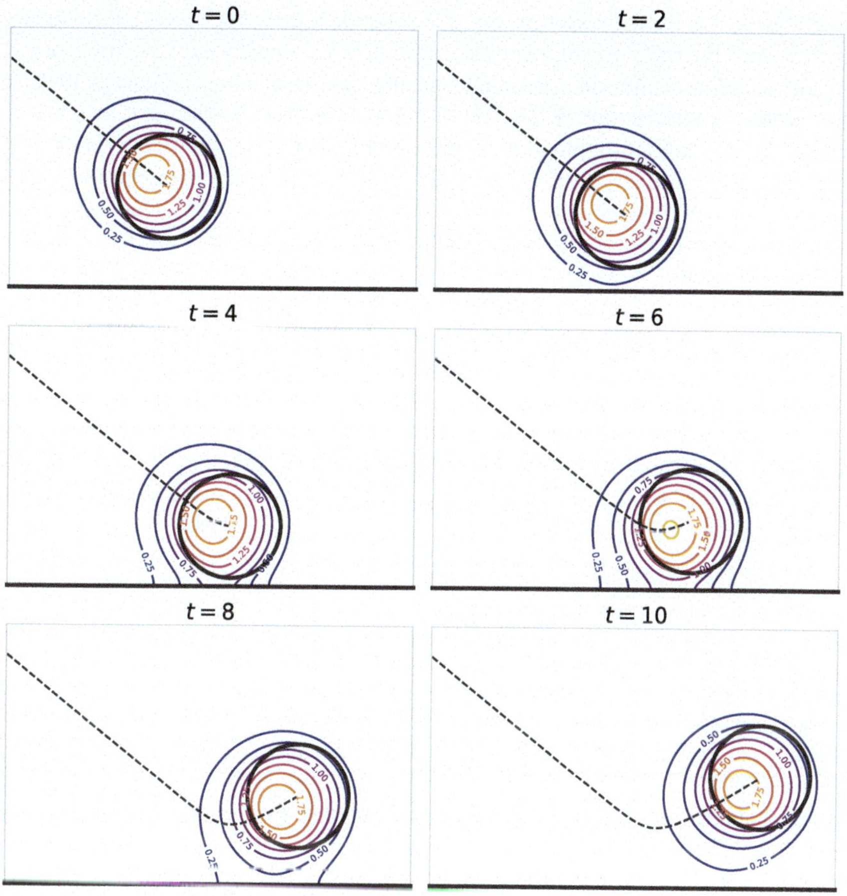

Fig. 1.13 Reflection in the moving boundary model. The black circle indicates the disk $B(\boldsymbol{p}(t), r_0)$, the dashed curve is the trajectory of the center, the colored curves are contour lines of $c(t, \boldsymbol{x})$, and the horizontal solid straight line is the boundary of the domain

3. After reflection, its speed recovers.
4. Reflection is nonspecular. In particular, the angle of reflection is greater than that of incidence.

To be more precise, we have found that there are parameters for which the MB model has the above properties, and we will focus on such properties throughout the present book. Although the approximately piecewise linear trajectory of the disk resulting from successive reflections reminds us of mathematical billiards, our 'equivalent of the billiard ball' does not obey the usual rule of completely elastic reflection. It is more like a soft, energy-dissipating ball bouncing off a wall, for which the coefficient of restitution is less than one, so that the ball travels more slowly after hitting the wall and thus moves at a more acute angle to the wall to preserve its momentum parallel to it. As we have seen in Fig. 1.13, the disk does not touch the boundary itself; instead the surface camphor concentration field interacts with the boundary. If the coefficient of restitution is less than one and there is no source of energy to drive the system, then the motion will eventually stop. However, the camphor disk has an engine to drive itself: it supplies camphor molecules to the water surface and thereby alters the water's surface tension. Unless this process is perfectly symmetrical, it will create a force on the camphor disk and thus enable it to recover a constant speed as we have seen in Fig. 1.10. The camphor disk is thus a kind of billiard ball in a dissipative system.

Since the angle of reflection differs from that of incidence, it is expected that quite different behavior to that of mathematical billiards will appear. For example, Fig. 1.14 shows a trajectory of the moving boundary model in a square domain. It seems that the disk is attracted to a square-shaped periodic orbit as time passes (a periodic orbit which is a limit of some solutions as $t \to \infty$ is called a limit cycle). In contrast, classical billiards in a square do not have any attractor [72]. For classical billiards, it is known that the behavior sensitively depends on the initial direction $\boldsymbol{v} = (v, w)$:

1. The orbit is periodic if w/v is rational, or
2. dense if w/v is irrational.

Although one may think that a square-shaped limit cycle is natural because the domain is square, other kinds of nontrivial orbits can appear when the domain is a rectangle. As shown in Fig. 1.15, very complicated periodic orbits may be observed.

Thus, our "billiard ball" exhibits quite different behavior from the mathematical billiard ball. However, it is difficult to prove the existence of a limit cycle of the moving boundary model. It is also difficult to simulate the moving boundary model so accurately that we can predict its asymptotic behavior. To understand this problem in more detail, we need a mathematical model that is more amenable to analysis. We

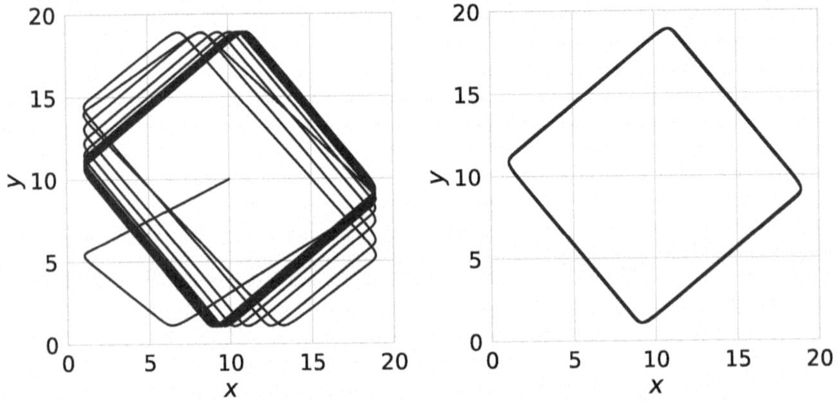

Fig. 1.14 Trajectories of the center of the disk in the moving boundary model on a square domain. Left: $0 \leq t \leq 10^4$, right: $9.5 \times 10^3 \leq t \leq 10^4$

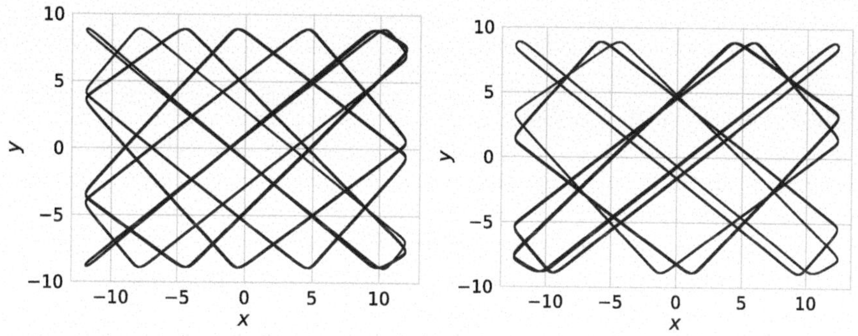

Fig. 1.15 Complicated periodic orbits for (1.11) on rectangular domains with $k = 1.3$ (left) and $k = 1.35$ (right)

next derive a model in which we regard the camphor disk as a point particle, assuming also that the motion of the disk is sufficiently slow.

We now focus on the interaction between two or more camphor disks in one system. To model this case, we introduce a moving boundary model for n disks. For $j = 1, 2, \ldots, n$, let $B_j(t)$ be the disk of radius $r_j > 0$ centered at $\boldsymbol{p}_j(t)$. The center $\boldsymbol{p}_j(t)$ of each disk is governed by

$$\dot{\boldsymbol{p}}_j = \frac{1}{\mu\pi r_j^2} \int_{\partial B_j(t)} \Gamma(c(t, \boldsymbol{x}))\boldsymbol{n} ds, \tag{1.4}$$

and the concentration $c(t, \boldsymbol{x})$ of camphor molecules on the water's surface is governed by

$$c_t = D\Delta c - \alpha c \chi_{\mathbb{R}^2 \setminus \cup B_j(t)} + a_0 \sum_{j=1}^{n} \chi_{B_j(t)}. \tag{1.5}$$

Equations (1.4)–(1.5) are the moving boundary model for multiple disks.

First, let us consider the motion of a single disk, which corresponds to (1.2)–(1.3), that is, (1.4)–(1.5) with $n = 1$. From the observation of Fig. 1.7, there exists a threshold $a_* > 0$: the disk eventually stops if $a_0 < a_*$, while it can keep moving if $a_0 > a_*$. Note that the motion of the disk is very slow when the purity parameter a_0 is sufficiently close to a_*.

The rest state is characterized by a stationary spot solution of the moving boundary model, while the moving state corresponds to a traveling spot solution. The stationary spot solution can be given by

$$c(t, \boldsymbol{x}) = S(\boldsymbol{x})$$

with a radially symmetric function $S(\boldsymbol{x})$. That is, there exists a function $\tilde{S} : \mathbb{R} \to \mathbb{R}$ and a constant b such that $S(\boldsymbol{x}) = S(|\boldsymbol{x}|)$ and

$$\lim_{r\to+\infty} \tilde{S}(r) = \frac{\exp(-\alpha r)}{\sqrt{r}} b, \tag{1.6}$$

where α is the sublimation rate of $c(t, \boldsymbol{x})$.

As we will see in Chap. 3 in detail, the linearized eigenvalues about the stationary spot have the following properties: zero is an eigenvalue and the corresponding eigenspace is two-dimensional for any choice of parameter values because of translation symmetry. In particular, the generalized eigenspace of the zero eigenvalue becomes four-dimensional at $a_0 = a_*$. Now, for $a_0 \approx a_*$, we approximate $c(t, \boldsymbol{x})$ by

$$c(t, \boldsymbol{x}) \approx S(\boldsymbol{x} - \boldsymbol{q}(t)) + v(t)\psi_1(\boldsymbol{x} - \boldsymbol{q}(t)) + w(t)\psi_2(\boldsymbol{x} - \boldsymbol{q}(t)), \tag{1.7}$$

where ψ_1 and ψ_2 are generalized eigenfunctions of zero eigenvalue. Truncating the equations for $\boldsymbol{q}(t) = (x(t), y(t))$ and $\boldsymbol{v}(t) = (v(t), w(t))$ at third order terms with respect to $\boldsymbol{v}$ and $a_0 - a_*$, we obtain the following ODEs:

$$\begin{cases} \dot{\boldsymbol{q}} = \boldsymbol{v} \\ \dot{\boldsymbol{v}} = (\delta - m_1|\boldsymbol{v}|^2)\boldsymbol{v}, \end{cases} \tag{1.8}$$

where δ is proportional to $a_0 - a_*$, and m_1 is a constant that appears in the process of derivation. We remark that $\boldsymbol{v}$ represents the deformation of the spot and that it corresponds to its velocity as can be seen from (1.8). Note that the shape of the camphor disk never changes but the concentration fields $c(t, \boldsymbol{x})$ can change.

Suppose that m_1 is positive. It is easy to solve (1.8). From $m_1 > 0$, it turns out that

$$\lim_{t\to+\infty} |\boldsymbol{v}(t)| = 0$$

if $\delta \leq 0$, and

$$\lim_{t\to+\infty} |\boldsymbol{v}(t)| = \sqrt{\frac{\delta}{m_1}}$$

if $\delta > 0$. A bifurcation diagram is shown in Fig. 1.16. Since the direction of motion is conserved for any $t > 0$, it is proved that the disk tends to uniform motion with a constant speed as time passes if the purity is above the threshold ($\delta > 0$).

Thus, under the assumption that the disk moves sufficiently slowly, we can regard it as a point particle and reduce the moving boundary model to ordinary differential equations that describe the center of the disk. In this sense, we can say that a camphor disk is a self-propelling particle. We call (1.8) a particle model.

From the viewpoint of infinite dimensional dynamical systems, the particle model is a vector field on a parameter-dependent center manifold, and a truncated normal form of a pitchfork bifurcation. By analyzing (1.8), one can prove that the stationary spot solution becomes unstable and that a stable traveling spot solution appears as a parameter varies. This forms one of the steps of mathematical analysis of the moving boundary model. Please see Chap. 3 for details.

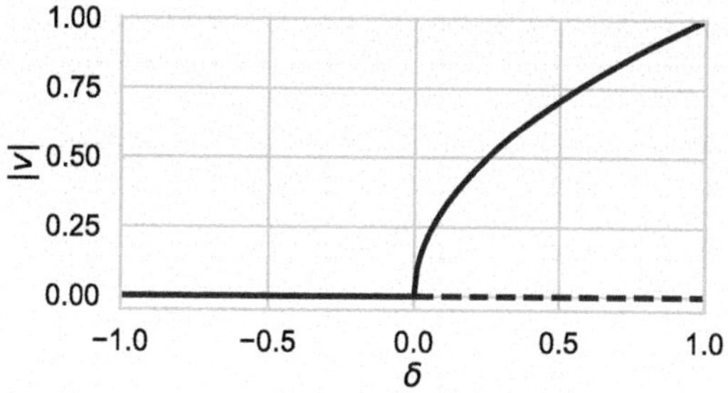

Fig. 1.16 A bifurcation diagram of (1.8) with $m_1 > 0$

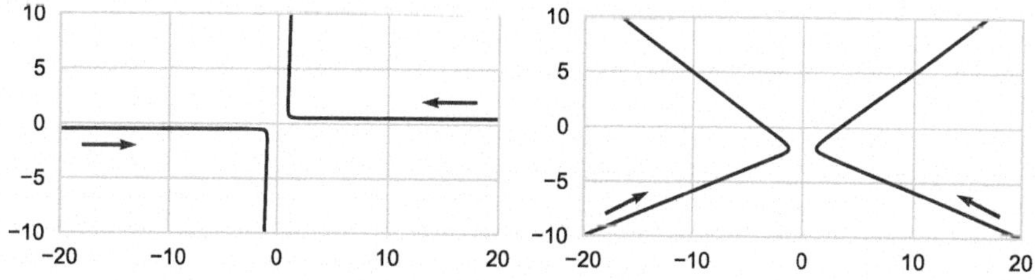

Fig. 1.17 Reflection of two particles according to (1.10) ($m_0 = m_1 = m_2 = \alpha = 1, \delta = 0.06$)

Next, let us consider interaction between two camphor disks (particles) $(\boldsymbol{q}_0, \boldsymbol{v}_0)$ and $(\boldsymbol{q}_1, \boldsymbol{v}_1)$. For (1.4)–(1.5) with $n = 2$ and $a_0 \approx a_*$, consider a solution of the form

$$c(t, \boldsymbol{x}) = S(\boldsymbol{x} - \boldsymbol{q}_0(t)) + S(\boldsymbol{x} - \boldsymbol{q}_1(t)) + \cdots . \tag{1.9}$$

It turns out that $(\boldsymbol{q}_0, \boldsymbol{v}_0)$ and $(\boldsymbol{q}_1, \boldsymbol{v}_1)$ are governed by

$$\begin{cases} \dot{\boldsymbol{q}}_j = \boldsymbol{v}_j - m_0 \dfrac{e^{-\alpha|\boldsymbol{d}_j|}}{\sqrt{|\boldsymbol{d}_j|}} \dfrac{\boldsymbol{d}_j}{|\boldsymbol{d}_j|}, \\ \dot{\boldsymbol{v}}_j = \left(\delta - m_1 |\boldsymbol{v}_j|^2\right) \boldsymbol{v}_j - m_2 \dfrac{e^{-\alpha|\boldsymbol{d}_j|}}{\sqrt{|\boldsymbol{d}_j|}} \dfrac{\boldsymbol{d}_j}{|\boldsymbol{d}_j|}, \end{cases} \quad (j = 0, 1), \tag{1.10}$$

where m_0, m_1, m_2 are positive constants, and $\boldsymbol{d}_0$ and $\boldsymbol{d}_1$ are defined by

$$\boldsymbol{d}_0 = \boldsymbol{q}_1 - \boldsymbol{q}_0, \quad \boldsymbol{d}_1 = \boldsymbol{q}_0 - \boldsymbol{q}_1.$$

The additional terms on the right hand sides are derived from the tail of the function S that behaves like (1.6). One may wonder why $\dot{\boldsymbol{q}}_j$ is not equal to $\boldsymbol{v}_j$. Please remember that (v, w) represents not velocity but deformation of the spot. Since m_0 and m_2 are positive, the interaction between two spots is repulsive. Indeed, one can observe the mutual rebound of two particles by simulation of (1.10), as shown in Fig. 1.17.

1.4 Particle Model on a Rectangle

Next, let us consider a mathematical model for a particle on a rectangular domain $[-kL, kL] \times [-L, L]$. Let (x, y) and (v, w) be the position and velocity of the particle. (More precisely, (v, w) is not the velocity but deformation of the spot from its stationary state.) Suppose that there are four virtual particles located symmetrically with respect to each edge of the rectangle. If they interact on the rectangle, then we have the particle model on the rectangle:

$$\begin{cases} \dot{x} = v + m_0(h(kL + x) - h(kL - x)), \\ \dot{v} = \left\{\delta - m_1(v^2 + w^2)\right\} v \\ \qquad + m_2(h(kL + x) - h(kL - x)), \\ \dot{y} = w + m_0(h(L + y) - h(L - y)), \\ \dot{w} = \left\{\delta - m_1(v^2 + w^2)\right\} w \\ \qquad + m_2(h(L + y) - h(L - y)), \end{cases} \tag{1.11}$$

where $h(x) = (2x)^{-\frac{1}{2}} e^{-2\alpha x}$. If the terms involving h were absent, then the equation would be nothing but (1.8). Therefore, if the particle is far away from the boundary, then the particle tends to uniform motion in a straight line. The h terms represent repulsive interactions with the boundary, which operate when the particle is near the boundary. Figure 1.18 illustrates an orbit on a square domain. The particle is attracted to a square limit cycle.

Figure 1.19 is an orbit diagram, where the aspect ratio k of the rectangle is treated as a parameter. This figure plots the values of x/kL when the orbit crosses $y = 0$ from negative to positive for various values of k. There are some intervals of k in which the points are very sparse, while there are values of k for which the points are

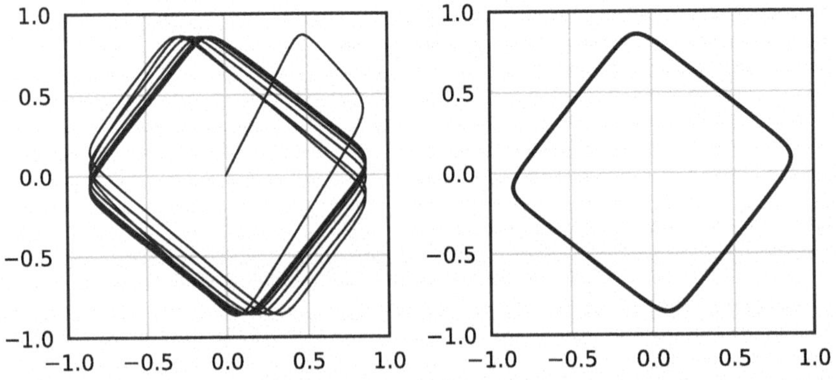

Fig. 1.18 An orbit of the particle for (1.11) on a square (left) and an attractor (right). The horizontal and vertical axes indicate x/L and y/L, respectively. Figure adapted from the author's previously published work [50]

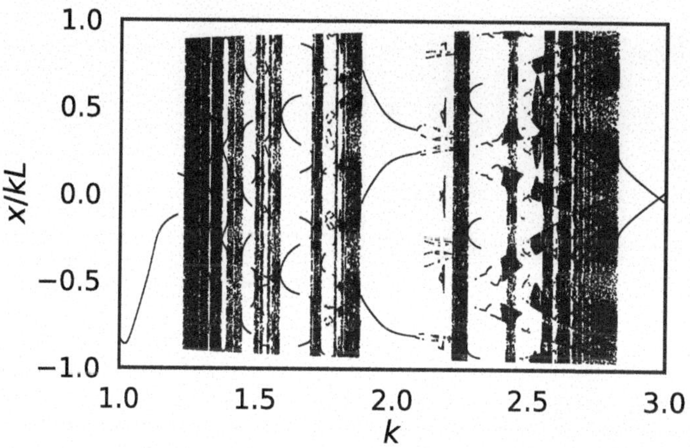

Fig. 1.19 An orbit diagram for (1.11). The horizontal and vertical axes are k and x/kL, respectively. Figure adapted from the author's previously published work [49]

dense. In the former parameter intervals, one can observe a periodic orbit. We call these parameter intervals windows, for the latter parameter values, one can observe a periodic orbit with very long period or other kinds of orbits. Figure 1.20 shows asymptotic orbits for $k = 1.3, 1.6, 1.85, 3.0$. According to numerically computed Lyapunov exponents, the orbits for $k = 1.6, 3.0$ are periodic, that for $k = 1.3$ is quasiperiodic, and that for $k = 1.85$ is chaotic [49]. Thus, the attractors depend on the aspect ratio k in a very complicated manner.

The intermittent change of attractors depending on the aspect ratio k can be explained in the following manner. We note that (1.11) is a coupled oscillator system consisting of two limit cycle oscillators (x, v) and (y, w), which bifurcate from the rest state as parameters vary. These two oscillators correspond to horizontal and vertical oscillations, respectively, and these are coupled through the nonlinear terms vw^2 and v^2w. The square-shaped cycle in the square domain may be considered as a synchronized state where the frequency ratio is $1:1$. The frequency ratio varies depending on the parameters, and the windows in Fig. 1.19 are caused by a phase-locking phenomenon. Indeed, bifurcation analysis around the rest state $(x, v, y, w) = (0, 0, 0, 0)$ and numerical computation located symmetrically suggest that an invariant torus bifurcates from horizontal oscillation as a secondary bifurcation. Complex limit cycles and quasiperiodic orbits are realized as dynamics on the invariant torus. In addition, the chaotic orbits appear in the parameter region where the invariant torus breaks down.

By local bifurcation theory and computer-aided analysis, it becomes possible to discuss limit cycles with small amplitudes and to explain the change of attractors as shown in Fig. 1.7. Is it possible to understand the asymptotic dynamics of billiard-like behavior resulting from reflection and uniform motion? Although the particle model is easier than the moving boundary model, there is no general method of proving

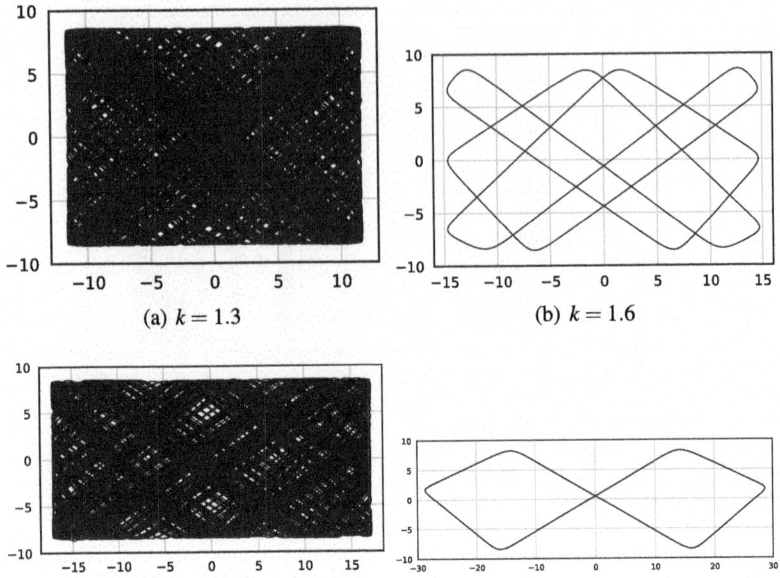

(a) $k = 1.3$ (b) $k = 1.6$

(c) $k = 1.85$ (d) $k = 3.0$

Fig. 1.20 Orbits of (1.11) on a rectangle. Figure adapted from the author's previously published work [49]

the existence of a limit cycle for a four-dimensional ODE. We therefore introduce another model to understand a square-shaped attractor in a square domain.

1.5 Discrete-Time Model

In order to derive another model, let us more closely observe the behavior of the particle model. Figure 1.21 shows square-shaped cycles for $L = 10, 25, 10^3$, where the $(x/L, y/L)$ is drawn. As L increases, the cycle gets close to the boundary. The particle looks as if it collides with the boundary for $L = 10^3$, although no actual collision ever occurs. Moreover, Fig. 1.22 shows the speed changing with time for $L = 10, 10^2, 10^3$. The speed decreases when reflection occurs, after which it recovers to its constant value when the particle is far away from the boundary. As the system size L increases, the relative time associated with reflection within a period becomes relatively small. That is, we can suppose that the particle maintains a constant speed if L is sufficiently large.

Taking the above observation into account, we may consider reflection and motion in a straight line in the limit $L \to \infty$, and scale down to the square domain $[-1, 1] \times [-1, 1]$. By this limit operation, we assume that the particle moves in a straight line and changes its direction after colliding with the boundary. If we know the reflection

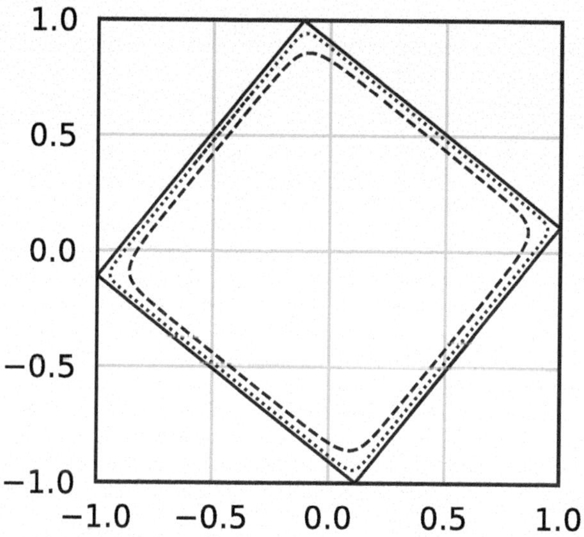

Fig. 1.21 Square-shaped limit cycles for (1.11) in square. The horizontal and vertical axes are x/L and y/L, respectively. The dashed, dotted, and solid lines indicate orbit for $L = 10, 25, 10^3$, respectively

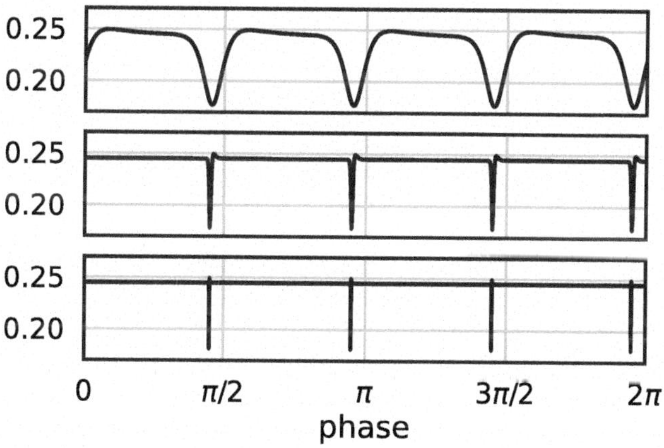

Fig. 1.22 The speed along a limit cycle for $L = 10, 10^2, 10^3$ from above. The horizontal axis indicates the phase

rule, the relation between the angles of incidence and reflection, we can determine the direction of motion when the reflection occurs and compute the reflection position and angle of reflection subsequently. We call this discrete-time dynamical system a discrete model.

Let us now consider the reflection rule of the particle. Suppose that there is a particle on the half plane $\{(x, y) \mid x > 0\}$. The y-axis is the boundary. The repulsive

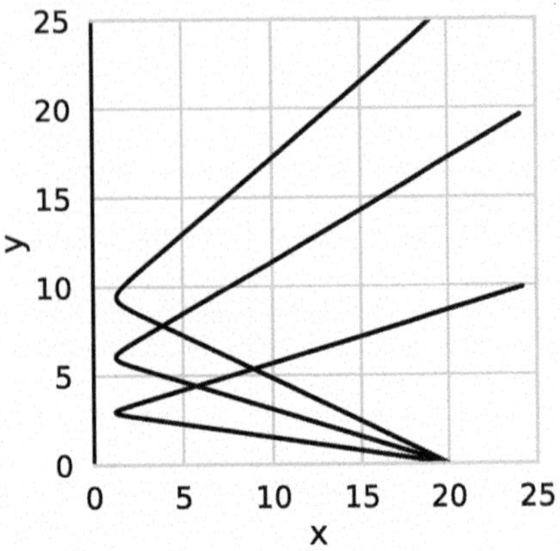

Fig. 1.23 Solution orbits of (1.12). Figure adapted from [15, Fig. 2(A)], used with permission of the American Institute of Mathematical Sciences, from Ei et al. [15]; permission conveyed through Copyright Clearance Center, Inc.

interaction with the boundary is equivalent to that from a particle which is situated symmetrically with respect to the y-axis. That is, the particle is governed by the following equation:

$$\begin{cases} \dot{x} = v + m_0 h(x), \\ \dot{v} = \left\{\delta - m_1(v^2 + w^2)\right\} v + m_2 h(x), \\ \dot{y} = w, \\ \dot{w} = \left\{\delta - m_1(v^2 + w^2)\right\} w. \end{cases} \tag{1.12}$$

We call (1.12) the particle model on the half plane. By numerical simulation of (1.12), we can observe reflection near the boundary as shown in Fig. 1.23. Furthermore, we can compute the reflection rule $\theta_{\mathrm{ref}} = F(\theta_{\mathrm{inc}})$ by computing trajectories for various angles of incidence, where θ_{inc} and θ_{ref} are angles of incidence and reflection, respectively. Here, we use F to represent the relationship between the angles, which differs from the previous use. Figure 1.24 shows reflection rules for some parameter values. Evidently, the angle of reflection is greater than that of incidence. In fact, it is possible to prove this property under the assumption that $m_0 = 0$[15]. It should be noted that $m_0 = 0$ is likely just a technical assumption. Although we believe that this assumption can be waived, that possibility is not yet proven in the case of $m_0 > 0$.

The existence of a square-shaped limit cycle in a square domain is proved in the following way. Assume that the particle visits all the edges in turn as shown in Fig. 1.25. The reflection position x and the angle of incidence θ is governed by a two-dimensional map

$$\mathscr{S} : \begin{pmatrix} x \\ \theta \end{pmatrix} \mapsto \begin{pmatrix} (1-x)\cot F(\theta) \\ \frac{\pi}{2} - F(\theta) \end{pmatrix}. \tag{1.13}$$

The square-shaped cycle is characterized by a point satisfying $(x_*, \theta_*) = \mathscr{S}(x_*, \theta_*)$, that is, fixed point (x_*, θ_*) of $\mathscr{S}$. The unique existence of the fixed point is guaranteed

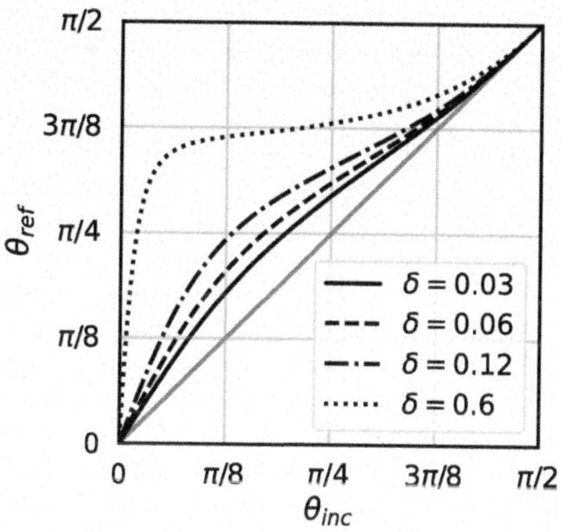

Fig. 1.24 Some reflection rules obtained from numerical simulation for (1.12) for some values of δ. Figure adapted from the author's previous work [50]

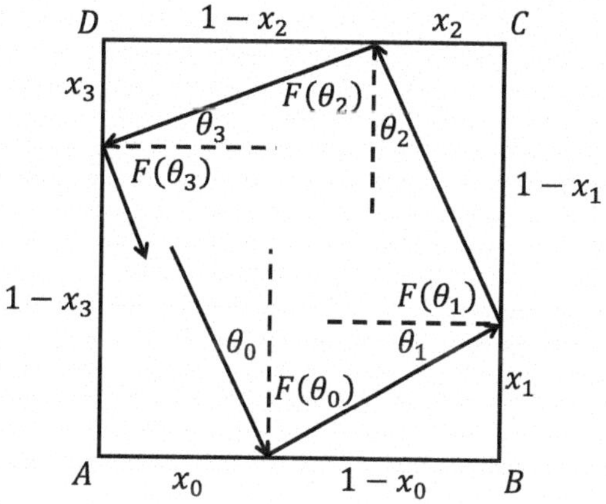

Fig. 1.25 Illustration of the discrete model. Figure adapted from [15, Fig. 8], used with permission of the American Institute of Mathematical Sciences, from Ei et al. [15]; permission conveyed through Copyright Clearance Center, Inc.

by the intermediate value theorem and the monotonicity of F. Its linear stability is determined by whether $|F'(\theta_*)| < 1$. In particular, orbits obtained by the reflection rule in Fig. 1.24 are linearly stable. Thus, we can prove the existence and stability of a square-shaped cycle by elementary mathematics.

Figure 1.26 shows an orbit diagram: the horizontal axis is the aspect ratio k and the vertical axis is arguments of complex numbers corresponding to the points of reflection. We can observe periodic orbits, such as rectangular cycles, figure-eight-shaped cycles, and so on. It is possible to prove the existence of such orbits for given k by elementary calculations. However, Fig. 1.26 appears more sparse than Fig. 1.16. Quasiperiodic or chaotic orbits are absent in Fig. 1.26. Although the discrete model approximates the particle model well for sufficiently large L, the disagreement between these models becomes large at mesoscopic scale, such as $L = 10$. It is necessary at this point to recall that the particle governed by the particle model interacts with all the edges of the domain. Thus, it is repulsed from two edges simultaneously when it is near a corner of the domain. However, reflection from only one edge is assumed in the discrete model. Two of the present authors proposed a

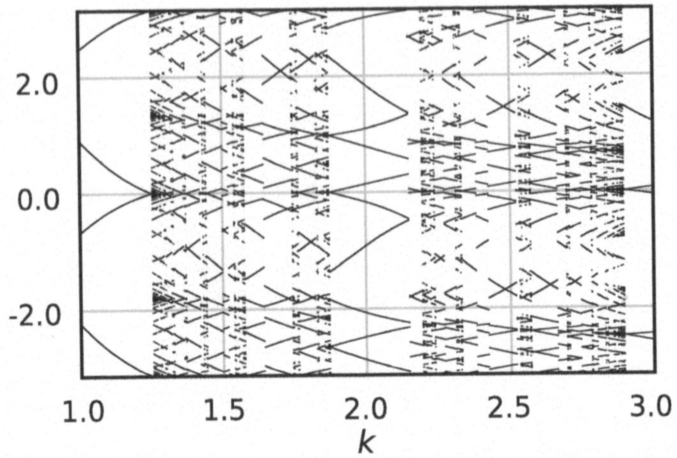

Fig. 1.26 Orbit diagram for the discrete model defined by (1.12) with $\delta = 0.06$. The horizontal and vertical axes indicate the aspect ratio and arguments of complex numbers corresponding to the reflection points. Figure adapted from the author's previous work [49]

modified discrete model in which the reflection rule depends on reflection positions, and they showed that quasiperiodic and chaotic orbits can appear in this modified model [49].

The discrete model is useful for proving the existence or nonexistence of a cycle that has a prescribed shape. However, it is not easy to develop further analysis because the discrete model is a two-dimensional dynamical system for reflection positions and angles of incidence. The modified model proposed in [49] is more complicated. To explain the change of attractor depending on the aspect ratio, the particle model was more useful than the discrete model.

In this chapter, a summary of the present book has been presented, and three mathematical models have been introduced. These models describe the motions of an individual but at different spatiotemporal scales. The MB model models a mechanism of self-propulsion. In the particle model, we regard the disk as a point particle. This assumption is valid provided that the disk moves sufficiently slowly, or, equivalently, a parameter is sufficiently close to the bifurcation point at which self-sustaining motion appears. The particle model does not model the mechanism of self-propulsion, but it describes how the disk moves and interacts with a boundary. In the discrete-time model, the domain size is assumed to be so large that the particle hits the boundary. Figure 1.27 illustrates relation among these models. These models will be discussed in detail throughout the chapters that follow.

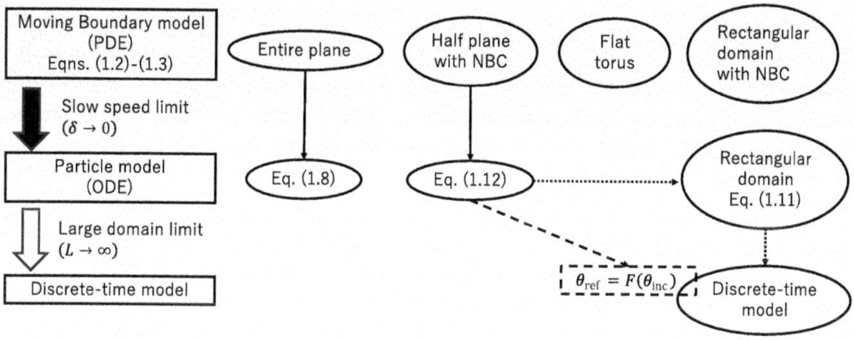

Fig. 1.27 Mathematical models in the present book. In the figure, NBC means the Neumann boundary condition. Each ellipse indicates a mathematical model, and each arrow connecting two ellipses means that the model at the arrow's head is derived from the model at the arrow's tail in some sense. Derivation of a particle model from a moving boundary model is justified by the theory in the next section

1.6 Some Remarks

According to [58, Sect. 1.2.1], a study on camphor motion was reported by Tomlinson in 1862 [79], and Lord Rayleigh published a paper on this topic in 1890 [75]. Several materials that exhibit self-propelled motion due to surface tension have been reported [76]. As far as the authors are aware, the modern study of camphor motion was initiated by Nakata et al. [57], who have reported experiments on two types of motion of camphor scrapping, translation and rotation. Nakata and Hayashima have shown that camphor scrapping can spontaneously change its motion between rotation and translation [55]. Nakata et al. have investigated rotating motion in a dumbbell-shaped region [56]. Hayashima et al. [26] have studied the oscillatory motion of a camphor disk in linear water channel with the aid of a mathematical model, numerical simulation, and laboratory experiments. Since, camphor motions have been studied by both experiments and mathematical models.

The mathematical model proposed by Hayashima et al. [26] also consists of equations of motion and a diffusion equation with sublimation and camphor supply together, where the effect of surface tension is described by the gradient of camphor concentration instead of a contour integral. Numerical simulation [53] suggests that it exhibits a subcritical drift bifurcation, depending on the parameter values. If camphor scrapping is not a disk, then rotational motion should be taken into account. See [36, 37, 59] for details of the model. Recently, a sophisticated phase-field model that describes the motion of self-propelled bodies driven by surface tension on the water surface has been proposed by Nagayama et al. [54]. Their model is not limited to a camphor solid but includes the case of a liquid droplet floating on the water surface. We have employed the moving boundary model for a camphor disk from [9].

Furuya [21] has studied a phase-field model by numerical simulation. They have found that a stationary spot loses its stability, and a traveling spot bifurcates as a parameter corresponding to the purity of the camphor disk increases. In addition, repulsive interactions between two traveling spots and reflection from a boundary have been observed. Kanda [32] has conducted experiments on the motions of camphor disks and has studied the phase-field model by numerical simulation. Matsumoto [47] has performed numerical simulations of the particle model on the entire plane and with a single wall to observe asymptotic motion in a straight line and reflection from a boundary. She has also considered the particle model on a square and two particles on a square.

First, the third author's research group considered camphor experiments of a concrete example of drift bifurcation. Later, as they studied the mathematical model, they noticed the relationship between the angles of incidence and reflection. Although they conducted some experiments to investigate the angle relations, more results were needed to develop precise hypotheses. We hope that chemists will validate the experiment described in this chapter in the future. Comparisons with mathematical models should also be made.

Chapter 2
Preliminaries

Abstract In this chapter, some basic concepts are introduced without proof. First, we introduce function spaces and infinite dimensional dynamical systems, which are helpful for understanding the discussion in Chap. 3. Second, we introduce finite dimensional continuous-time dynamical systems defined by ODEs, which are used in Chap. 4 to study particle models. Finally, we introduce discrete-time dynamical systems defined by maps, which are used in Chap. 5. See [27] for details of the materials in Sect. 2.1. If a reader is not yet initiated in functional analysis, [7] will prove useful, which is an introduction to functional analysis with applications to PDE. See [24, 43] for a description of dynamical systems and bifurcations, See [25, 82] for the center manifold theory.

2.1 Function Spaces and Infinite Dimensional Dynamical Systems

L^p Space

Let $n \in \mathbb{N}$ and $p \geq 1$. Let $\Omega \subseteq \mathbb{R}^n$ be an open region. We denote the space of pth power integrable functions by $L^p(\Omega)$. The L^p space equipped with the norm

$$\|f\|_p = \left(\int_\Omega |f(x)|^p dx \right)^{\frac{1}{p}}$$

is a Banach space. In particular, $L^2(\Omega)$ is a Hilbert space with an inner product

$$(f, g)_{L^2} = \int_\Omega f(x)g(x)dx.$$

T. Miyaji et al., *A Billiard Problem in Nonlinear Dissipative Systems*,
Surveys and Tutorials in the Applied Mathematical Sciences 18,
https://doi.org/10.1007/978-981-95-7561-9_2

We denote the Sobolev space of L^p functions, which have all derivatives up to kth order by $W^{k,p}(\Omega)$. In particular, we denote $H^k(\Omega) = W^{k,2}(\Omega)$. $H^k(\Omega)$ is a Hilbert space with the inner product

$$(f, g)_{H^k} = \sum_{j=0}^{k} \int_{\Omega} \frac{\partial^j f}{\partial x^j}(x) \frac{\partial^j g}{\partial x^j}(x) dx.$$

Spectrum

For a Banach space X, the set of bounded linear operators on X is denoted by $\mathscr{L}(X)$. Let A be a linear operator $A : D(A) \subset X \to X$, where $D(A)$ is the domain of A. The resolvent set of A is the set of complex numbers defined by

$$\rho(A) = \left\{\lambda \in \mathbb{C} \mid (\lambda - A) \text{ is invertible and } (\lambda - A)^{-1} \in \mathscr{L}(X)\right\}.$$

The complement set $\mathbb{C} \setminus \rho(A)$ is referred to as the spectrum of A, and it is denoted by $\sigma(A)$. A complex number λ is called an eigenvalue of A if $\ker(\lambda - A) \neq \{0\}$. We denote the set of all eigenvalues of A by $\sigma_{\mathrm{p}}(A)$. Any complex number in $\sigma(A) \setminus \sigma_{\mathrm{p}}(A)$ is called an essential spectrum of A. The set of all essential spectra is denoted by $\sigma_{\mathrm{ess}}(A)$. We denote $\mathrm{Re}\sigma(A) > 0$ if any spectrum λ of A has a positive real part.

A linear operator A in a Banach space X is said to be sectorial if and only if it is closed densely defined operator and there exist $\phi \in (0, \pi/2)$, $M \geq 1$, and $a \in \mathbb{R}$ such that

$$S_{a,\phi} = \{\lambda \mid \phi \leq |\arg(\lambda - a)| \leq \pi, \lambda \neq a\} \subset \rho(A),$$

and for any $\lambda \in S_{a,\phi}$

$$\|(\lambda - A)^{-1}\| \leq \frac{M}{|\lambda - a|}$$

holds.

For a sectorial operator A with $\mathrm{Re}\sigma(A) > 0$ and $\alpha > 0$, let $A^{-\alpha}$ be defined by

$$A^{-\alpha} = \frac{1}{\Gamma(\alpha)} \int_0^\infty t^{\alpha-1} e^{-tA} dt,$$

where Γ is the gamma function. $A^{-\alpha}$ is a bounded linear operator on X. For $\alpha > 0$, A^α is defined as the inverse of $A^{-\alpha}$ and its domain $D(A^\alpha)$ is the range of $A^{-\alpha}$. A^0 is defined as the identity operator I on X. When $\mathrm{Re}\sigma(A) > 0$ does not holds, choosing $a \in \mathbb{R}$ so that $\mathrm{Re}\sigma(A + aI) > 0$, define the space X^α as the domain of the linear operator $(A + aI)^\alpha$, that is, $X^\alpha = D((A + aI)^\alpha)$.

Semiflow

Let V be an open subset of X^α for some $0 \leq \alpha < 1$, and let $f : V \to X$ be a locally Lipschitz function. Let $u(t; u_0)$ be the solution of the initial value problem

$$\frac{du}{dt} + Au = f(u), \quad u(0; u_0) = u_0, \tag{2.1}$$

and let $S(t)$ be a map from the initial value u_0 to the solution $u(t; u_0)$ at time $t \geq 0$. Suppose that there exists a closed subset $C \subset V$ such that if $u_0 \in C$, then $S(t)u_0 = u(t; u_0) \in C$ holds for any $t \geq 0$. The family of maps $\{S(t)\}_{t\geq 0}$ defines a semiflow in C, that is, it satisfies the following four conditions [27, Definition 4.1.1]:

1. For each $t \geq 0$, $S(t)$ is continuous from C to C.
2. For each $u \in C$, $t \mapsto S(t)u$ is continuous.
3. $S(0)$ is the identity map on C.
4. For any $u \in C$ and $t, \tau \geq 0$,

$$S(t)(S(\tau)u) = S(t + \tau)u.$$

Stability

For $u \in C$, the set $\mathcal{O}^+(u) = \{S(t)u \mid t \geq 0\}$ is called the positive orbit of u. The point $u \in C$ is called an equilibrium if and only if $\mathcal{O}^+(u) = \{u\}$. An orbit $\mathcal{O}^+(u)$ is said to be orbitally stable if and only if for any $\varepsilon > 0$ there exists $\delta > 0$ such that for all $t \geq 0$ $\mathrm{dist}(u, v) < \delta$ implies $\mathrm{dist}(\mathcal{O}^+(u), S(t)v) < \varepsilon$. An orbit is said to be unstable if it is not stable. An orbit is said to be uniformly asymptotically stable if it is stable and there is a neighborhood $U \subset C$ such that

$$\lim_{t\to+\infty} \mathrm{dist}(S(t)v, \mathcal{O}^+(u)) = 0$$

uniformly for $v \in U$.

The following theorem, which is a summary of Theorems 5.1.1 and 5.1.3 of [27], provides a criterion for the stability of an equilibrium.

Theorem 2.1 *Suppose that x_0 is an equilibrium of* (2.1) *with*

$$f(x_0 + z) = f(x_0) + Bz + g(z),$$

where $\|g(z)\| = o(\|z\|)$ as $\|z\| \to 0$.

1. *If the spectrum of $L = -A + B$ lies in $\{Re\,\lambda \mid < -\beta\}$ for some $\beta > 0$, then the equilibrium x_0 is asymptotically stable in X^α.*
2. *If $\sigma(L) \cap \{Re\,\lambda > 0\} \neq \emptyset$, then the equilibrium x_0 is unstable.*

Center Manifold

The following theorem indicates that there exists an invariant manifold near an equilibrium whose linearized operator possesses zero as an eigenvalue. This allows us to reduce an infinite dimensional dynamical system to a finite dimensional one. See

[27, 82] for details of the center manifold theory in infinite dimensional systems. We cite Theorem 6.2.1 of [27]:

Theorem 2.2 *Suppose A is sectorial in X, $0 \leq \alpha < 1$, U is a neighborhood of the origin in X^α and $h : U \to X$ is continuously differentiable with $h(0) = 0$. Assume $L = A - h'(0)$ has $Re\,\sigma(L) \geq 0$ and $\sigma(L) \cap \{Re\,\lambda = 0\}$ is a spectral set. (If A has compact resolvent, this is a finite collection of eigenvalues.) We study the stability of the origin for $dx/dt + Ax = h(x)$.*

Let $X = X_1 \oplus X_2$ be the corresponding decomposition into L-invariant subspaces with $Re\,\sigma(L_1) = 0$, $Re\,\sigma(L_2) > 0$, $L_j = L|_{X_j}$ $(j = 1, 2)$. There exists a Lipschitzian local invariant manifold $S = \{x = x_1 + \phi(x_1) \mid x_1 \in x_1, \|x_1\| \leq r\}$, tangent to X_1 at the origin with $\phi(x_1) \in X_2$. The flow in S may be represented by the ordinary differential equation in X_1

$$\frac{dx_1}{dt} + L_1 x_1 = E_1 g(x_1 + \phi(x_1))$$

where $g(x) = h(x) - h'(0)x$ and E_1 is the projection of X onto X_1 (along X_2).

If the origin (in X_1) is asymptotically stable for the flow in S, the origin is asymptotically stable in X^α; if the origin is unstable for the flow in S, then it is unstable in X^α.

2.2 Continuous-Time Dynamical Systems Defined by ODEs

Flow

For a C^1 vector field $f : \mathbb{R}^n \to \mathbb{R}^n$, consider an ordinary differential equation

$$\dot{\boldsymbol{x}} = f(\boldsymbol{x}) \tag{2.2}$$

with the initial condition

$$\boldsymbol{x}(0) = \boldsymbol{\xi}. \tag{2.3}$$

By the fundamental theorem of ODEs, there exists a solution $\varphi(t, \boldsymbol{\xi})$ of (2.2)–(2.3). We denote $\varphi^t(\boldsymbol{\xi}) = \varphi(t, \boldsymbol{\xi})$. Suppose that the solution is defined for any $t \in \mathbb{R}$. The family $\varphi = \{\varphi^t\}_{t\in\mathbb{R}}$ defines a flow on $\mathbb{R}^n$, that is, it satisfies the following properties:

1. φ^0 is the identity map on $\mathbb{R}^n$.
2. For any $t, s \in \mathbb{R}$, $\varphi^{t+s} = \varphi^t \circ \varphi^s$.

Definition 2.3 The orbit of $\boldsymbol{\xi} \in \mathbb{R}^n$ by the flow φ is defined by

$$\mathcal{O}(\boldsymbol{\xi}) = \left\{\varphi^t(\boldsymbol{\xi}) \mid t \in \mathbb{R}\right\}.$$

The positive orbit of $\boldsymbol{\xi} \in \mathbb{R}^n$ by the flow φ is defined by

$$\mathscr{O}^+(\boldsymbol{\xi}) = \left\{\varphi^t(\boldsymbol{\xi}) \mid t \geq 0\right\}.$$

Definition 2.4 Let φ be a flow.

1. A point $\boldsymbol{p} \in \mathbb{R}^n$ is called an equilibrium of φ if and only if $\mathscr{O}(\boldsymbol{p}) = \{\boldsymbol{p}\}$.
2. A point $\boldsymbol{q} \in \mathbb{R}^n$ is called a periodic point of φ if and only if there exists $T > 0$ such that $\varphi^T(\boldsymbol{q}) = \boldsymbol{q}$ and there exists $t \in (0, T)$ such that $\varphi^t(\boldsymbol{q}) \neq \boldsymbol{q}$. The orbit of a periodic point is called a periodic orbit.

Definition 2.5 Let φ be a flow on $\mathbb{R}^n$, and let $S \subset \mathbb{R}^n$.

1. S is said to be invariant if $\mathscr{O}(\boldsymbol{x}) \subset S$ for any $\boldsymbol{x} \in S$.
2. S is said to be positively invariant if $\mathscr{O}^+(\boldsymbol{x}) \subset S$ for any $\boldsymbol{x} \in S$.

Definition 2.6 Let φ be a flow on $\mathbb{R}^n$.

1. $\boldsymbol{y} \in \mathbb{R}^n$ is called an ω-limit point of $\boldsymbol{x} \in \mathbb{R}^n$ if and only if there exists a monotone increasing sequence $\{t_k\}_{k=0}^{\infty}$ such that

$$\lim_{k\to+\infty} \varphi^{t_k}(\boldsymbol{x}) = \boldsymbol{y}, \quad \lim_{k\to+\infty} t_k = \infty.$$

   The set of all ω-limit points of $\boldsymbol{x}$ is called an ω-limit set of $\boldsymbol{x}$, and it is denoted by $\omega(\boldsymbol{x})$.
2. $\boldsymbol{y} \in \mathbb{R}^n$ is called an α-limit point of $\boldsymbol{y} \in \mathbb{R}^n$ if and only if there exists a monotone increasing sequence $\{t_k\}_{k=0}^{\infty}$ such that

$$\lim_{k\to+\infty} \varphi^{-t_k}(\boldsymbol{x}) = \boldsymbol{y}, \quad \lim_{k\to+\infty} t_k = \infty.$$

   The set of all α-limit points of $\boldsymbol{x}$ is called an α-limit set of $\boldsymbol{x}$, and it is denoted by $\omega(\boldsymbol{x})$.

Definition 2.7 A simple closed curve $\gamma \subset \mathbb{R}^n$ is called a limit cycle of a flow φ if and only if γ is a periodic orbit of φ and $\gamma = \omega(\boldsymbol{x})$ for some $\boldsymbol{x} \notin \gamma$.

Theorem 2.8 (Poincaré–Bendixson) *Let φ be a flow in an open region $X \subseteq \mathbb{R}^2$, and let $\boldsymbol{x} \in X$. If $\mathscr{O}^+(\boldsymbol{x})$ is contained in a compact subset of X and $\omega(\boldsymbol{x})$ contains no equilibrium, then $\omega(\boldsymbol{x})$ is a periodic orbit.*

Stability of Equilibrium of Flow

Definition 2.9
1. An equilibrium $\boldsymbol{p}$ of a flow φ is said to be stable if and only if for any $\varepsilon > 0$ there exists a $\delta > 0$ such that $\mathscr{O}^+(\boldsymbol{\xi})$ is contained in the ε-neighborhood of $\boldsymbol{p}$ for any $\boldsymbol{\xi}$ in the δ-neighborhood of $\boldsymbol{p}$.
2. An equilibrium is said to be unstable if it is not stable.

3. An equilibrium $\boldsymbol{p}$ is said to be asymptotically stable if it is stable and there exists $\eta > 0$ such that
$$\lim_{t\to+\infty} \varphi^t(\xi) = \boldsymbol{p}$$
for any ξ with $|\xi - \boldsymbol{p}| < \eta$.

Definition 2.10 Let $\boldsymbol{p}$ and $\boldsymbol{q}$ be equilibria.

1. An orbit $\mathcal{O}(\boldsymbol{x})$ is a heteroclinic orbit from $\boldsymbol{p}$ to $\boldsymbol{q}$ if
$$\lim_{t\to-\infty} \varphi^t(\boldsymbol{x}) = \boldsymbol{p}, \quad \lim_{t\to+\infty} \varphi^t(\boldsymbol{x}) = \boldsymbol{q}.$$
A solution corresponding to the heteroclinic orbit is called a heteroclinic solution of (2.2).
2. An orbit $\mathcal{O}(\boldsymbol{x})$ is a homoclinic orbit to $\boldsymbol{p}$ if $\mathcal{O}(\boldsymbol{x}) \neq \{\boldsymbol{p}\}$ and
$$\lim_{t\to\pm\infty} \varphi^t(\boldsymbol{x}) = \boldsymbol{p}$$
A solution corresponding to the homoclinic orbit is called a homoclinic solution of (2.2).

We recall some stability criteria for equilibria.

Definition 2.11 Let $\boldsymbol{p}$ be an equilibrium of a flow φ. Let $\Omega \subseteq \mathbb{R}^n$ be a neighborhood of $\boldsymbol{p}$. A function $V : \Omega \to \mathbb{R}$ is called a Lyapunov function if it satisfies the following conditions:

1. V is continuous,
2. $V(\boldsymbol{p}) = 0$ and $V(\boldsymbol{p}) < V(\boldsymbol{x})$ for any $\boldsymbol{x} \in \Omega \setminus \{\boldsymbol{p}\}$,
3. $V(\varphi^t(\boldsymbol{x}))$ is strictly monotone decreasing with respect to t for any $\boldsymbol{x} \in \Omega \setminus \{\boldsymbol{p}\}$.

Theorem 2.12 *Let $\boldsymbol{p}$ be an equilibrium of a flow φ. If there exists a Lyapunov function, then $\boldsymbol{p}$ is asymptotically stable.*

Theorem 2.13 *Let $\boldsymbol{p}$ be an equilibrium of the flow defined by* (2.2). *Let $A = f'(\boldsymbol{p})$ be the Jacobian matrix of f about $\boldsymbol{p}$.*

1. *If all eigenvalues of A have a negative real part, then $\boldsymbol{p}$ is asymptotically stable.*
2. *If A has at least one eigenvalue with a positive real part, then $\boldsymbol{p}$ is unstable.*

Invariant Manifolds

Definition 2.14 Let $\boldsymbol{p} \in \mathbb{R}^n$ be an equilibrium of the flow φ. The local stable manifold $W^s_{\text{loc}}(\boldsymbol{p})$ and the local unstable manifold $W^u_{\text{loc}}(\boldsymbol{p})$ of $\boldsymbol{p}$ are defined by
$$W^s_{\text{loc}}(\boldsymbol{p}) = \left\{\boldsymbol{x} \in U \mid \lim_{t\to\infty} \varphi^t(\boldsymbol{x}) = \boldsymbol{p} \text{ and } \varphi^t(\boldsymbol{x}) \in U \text{ for any } t \geq 0\right\},$$
$$W^u_{\text{loc}}(\boldsymbol{p}) = \left\{\boldsymbol{x} \in U \mid \lim_{t\to\infty} \varphi^{-t}(\boldsymbol{x}) = \boldsymbol{p} \text{ and } \varphi^t(\boldsymbol{x}) \in U \text{ for any } t \leq 0\right\},$$

where U is some neighborhood of $\boldsymbol{p}$. The stable and unstable manifolds of $\boldsymbol{p}$ are defined as follows:

$$W^s(\boldsymbol{p}) = \bigcup_{t\geq 0} \varphi^{-t}(W^s_{\rm loc}(\boldsymbol{p})) \quad W^u(\boldsymbol{p}) = \bigcup_{t\geq 0} \varphi^{t}(W^u_{\rm loc}(\boldsymbol{p})),$$

respectively.

Theorem 2.15 *([24, Theorem 3.2.1]) Let f be a C^r vector field on $\mathbb{R}^n$ vanishing at the origin ($f(0) = 0$) and let $A = Df(0)$. Divide the spectrum into three parts, $\sigma_s, \sigma_c, \sigma_u$ with*

$$Re\,\lambda \begin{cases} < 0 & if\, \lambda \in \sigma_s, \\ = 0 & if\, \lambda \in \sigma_c, \\ > 0 & if\, \lambda \in \sigma_u. \end{cases}$$

Let the (generalized) eigenspaces of σ_s, σ_c, and σ_u be E^s, E^c, and E^u, respectively. Then there exist C^r stable and unstable invariant manifolds W^s and W^u tangent to E^s and E^u at 0 and a C^{r-1} center manifold W^c tangent to E^c at 0. The manifolds W^u, W^s, and W^c are all invariant for the flow of f. The stable and unstable manifolds are unique, but W^c need not be.

Definition 2.16 An equilibrium $\boldsymbol{p}$ of a vector field f is said to be hyperbolic if $f'(\boldsymbol{p})$ has no eigenvalue on the imaginary axis. An equilibrium is said to be nonhyperbolic if it is not hyperbolic.

Hopf Bifurcation

Let f be a smooth function $f : \mathbb{R}^n \times \mathbb{R} \to \mathbb{R}^n$. Consider a family of ODEs

$$\dot{\boldsymbol{x}} = f(\boldsymbol{x}, \alpha). \tag{2.4}$$

Assume that the following conditions are satisfied:

1. $f(\boldsymbol{0}, 0) = \boldsymbol{0}$.
2. $A = D_x f(\boldsymbol{0}, 0)$ has a pair of eigenvalues $\pm i\omega_0 (\omega_0 > 0)$ and no other eigenvalues with zero real parts.

According to the implicit function theorem, there exists a function $\boldsymbol{x}_0(\alpha)$ $f(\boldsymbol{x}_0(\alpha), \alpha) = \boldsymbol{0}$ for sufficiently small $|\alpha|$ and $\boldsymbol{x}_0(0) = \boldsymbol{0}$. We denote the Taylor expansion of $f(\boldsymbol{x}, 0)$ up to cubic terms as follows:

$$f(\boldsymbol{x}, 0) = A\boldsymbol{x} + \frac{1}{2}B(\boldsymbol{x}, \boldsymbol{x}) + \frac{1}{6}C(\boldsymbol{x}, \boldsymbol{x}, \boldsymbol{x}) + O(|\boldsymbol{x}|^4).$$

Let $\boldsymbol{q}, \boldsymbol{p} \subset \mathbb{C}^n$ be nonzero vectors satisfying

$$A\boldsymbol{q} = i\omega_0\boldsymbol{q}, \quad A^T\boldsymbol{p} = -i\omega_0\boldsymbol{p},$$

with $\langle \boldsymbol{p}, \boldsymbol{q} \rangle = 1$. Here, $\langle \cdot, \cdot \rangle$ is the standard inner product on $\mathbb{C}^n$. Define the following two quantities:

$$\sigma'(0) = \mathrm{Re}\langle \boldsymbol{p}, D_{x\alpha} f(\boldsymbol{0}, 0)\boldsymbol{q} \rangle,$$
$$a_1 = \langle \boldsymbol{p}, C(\boldsymbol{q}, \boldsymbol{q}, \overline{\boldsymbol{q}}) - 2B(\boldsymbol{q}, A^{-1}B(\boldsymbol{q}, \overline{\boldsymbol{q}})) + B(\overline{\boldsymbol{q}}, (2i\omega_0 I_n - A)^{-1}B(\boldsymbol{q}, \boldsymbol{q})) \rangle.$$

Here, $\sigma(\alpha)$ is the real part of the critical eigenvalue $\lambda(\alpha)$ with $\lambda(0) = i\omega_0(\sigma(0) = 0)$ and $\sigma'(0)$ is its derivative at $\alpha = 0$.

The following theorem summarizes the Hopf bifurcation.

Theorem 2.17 *Under the settings above, the following holds true.*

1. *If* $\sigma'(0)a_1 < 0$, *then* (2.4) *has a limit cycle in a neighborhood of the origin for sufficiently small* $\alpha > 0$.
2. *If* $\sigma'(0)a_1 > 0$, *then* (2.4) *has a limit cycle in a neighborhood of the origin for sufficiently small* $\alpha < 0$.

If $a_1 < 0$, *then the limit cycle is stable. If* $a_1 > 0$, *then it is unstable.*

Normal Form

See [25] for details of normal form theory.

Consider a parameter-dependent system:

$$\dot{\boldsymbol{x}} = A\boldsymbol{x} + f(\boldsymbol{x}, \alpha). \tag{2.5}$$

Assume that A and f are as follows:

1. A is a $n \times n$ matrix.
2. For some $k \geq 2$, there exist neighborhoods $V_x \subset \mathbb{R}^n$ and $V_\alpha \subset \mathbb{R}^m$ of $\boldsymbol{0}$ such that $f \in C^k(V_x \times V_\alpha, \mathbb{R}^n)$ and $f(\boldsymbol{0}, \boldsymbol{0}) = \boldsymbol{0}$, $D_{xf}(\boldsymbol{0}, \boldsymbol{0}) = O$.

Theorem 2.18 *([25, Theorem 2.2 of Chap. 3]) Under the assumptions above, for any positive integer* p, $2 \leq p \leq k$, *there exist neighborhoods* V_1 *and* V_2 *of* $\boldsymbol{0}$ *in* $\mathbb{R}^n$ *and* $\mathbb{R}^m$, *respectively, such that for any* $\mu \in V_2$, *there is a polynomial* $\Phi_\mu : \mathbb{R}^n \to \mathbb{R}^n$ *of degree* p *with the following properties:*

(i) The coefficients of the monomials of degree q *in* Φ_μ *are functions of* μ *of class* C^{k-q}, *and*

$$\Phi_0(\boldsymbol{0}) = \boldsymbol{0}, \quad D_x\Phi_0(\boldsymbol{0}) = O.$$

(ii) For $\boldsymbol{y} \in V_1$, *the polynomial change of the variable*

$$\boldsymbol{x} = \boldsymbol{y} + \Phi_\mu(\boldsymbol{y})$$

transforms equation (2.5) *into the "normal form"*

$$\dot{y} = Ay + N_\mu(y) + \rho(y, \mu)$$

and the following properties hold:

(a) *For any* $\mu \in V_2$, N_μ *is a polynomial* $\mathbb{R}^n \to \mathbb{R}^n$ *of degree p, with coefficients depending upon* μ, *such that the coefficients of the monomials of degree q are of class* C^{k-q}, *and*

$$N_0(\mathbf{0}) = \mathbf{0}, \quad D_y N_0(\mathbf{0}) = O.$$

(b) *The equality*

$$N_\mu(e^{tA^*}y) - e^{tA^*}N_\mu(y)$$

holds for all $(t, y) \in \mathbb{R} \times \mathbb{R}^n$ *and* $\mu \in V_2$.

(c) *The map* ρ *belongs to* $C^k(V_1 \times V_2, \mathbb{R}^n)$, *and* $\rho(y, \mu) = O(|y|^p)$ *for all* $\mu \in V_2$.

Theorem 2.19 *([25, Theorem 3.2 of Chap. 3]) Under the assumptions of Theorem 2.18, further assume that there exists an isometry* $T \in \mathscr{L}(\mathbb{R}^n)$ *which commutes with* (2.5),

$$TLx = LTx, \quad Tf(x) = f(Tx)$$

for all $x \in \mathbb{R}^n$. *Then the polynomials* Φ_μ *and* N_μ *in Theorem 2.18 commutes with* T.

Poincaré Map

Let $p \in \mathbb{R}^n$ be a periodic point of the flow φ defined by (2.2). Let $T > 0$ be the minimal period of p, and let $\gamma = \mathscr{O}(p)$. One can study the stability and bifurcations of the periodic orbit in the following manner.

Let Σ be a hypersurface

$$\Sigma = \{x \in \mathbb{R}^n \mid \langle a, x - p\rangle = 0\}$$

where $a \in \mathbb{R}^n \setminus \{\mathbf{0}\}$. Suppose that $\langle a, f(p)\rangle \neq 0$ holds. As $p \in \Sigma$ is a periodic point, for $x \in \Sigma$ sufficiently close to p, there exists a $\tau = \tau(x) > 0$ such that $\varphi^{\tau(x)}(x) \in \Sigma$ and $\tau(p) = T$. We can define a map

$$\tilde{P} : \Sigma \ni x \mapsto \varphi^{\tau(x)}(x) \in \Sigma$$

in a sufficiently small neighborhood of p. As Σ is an $n-1$ dimensional hyperplane, we can regard $\tilde{P}$ as a map P on $\mathbb{R}^{n-1}$ with $P(\mathbf{0}) = \mathbf{0}$. This is called a Poincaré map associated with the periodic orbit γ. The stability of γ is equivalent to one of the fixed point $\mathbf{0} \in \mathbb{R}^{n-1}$ for P. Thus, the stability and bifurcation problems of periodic orbits are reduced to those in discrete dynamical systems.

2.3 Discrete-Time Dynamical Systems Defined by Maps

Stability of a Fixed Point of a Map

Definition 2.20 Let f be a map $\mathbb{R}^N \to \mathbb{R}^N$. A positive orbit of $\boldsymbol{x} \in \mathbb{R}^n$ under f is the set

$$\mathscr{O}^+(\boldsymbol{x}) = \left\{ f^n(\boldsymbol{x}) \mid n = 0, 1, 2, \ldots \right\}$$

Definition 2.21 Let f be a map $\mathbb{R}^n \to \mathbb{R}^n$.

1. A point $\boldsymbol{p} \in \mathbb{R}^n$ is said to be a fixed point of f if and only if $f(\boldsymbol{p}) = \boldsymbol{p}$ holds.
2. A point $\boldsymbol{q} \in \mathbb{R}^n$ is said to be a periodic point of f with period m if and only if $f^m(\boldsymbol{p}) = \boldsymbol{p}$ for $m \in \mathbb{N}$ holds.

Definition 2.22 Let $f : \mathbb{R}^n \to \mathbb{R}^n$ be a continuous map.

1. A fixed point $\boldsymbol{p}$ of f is said to be stable if and only if for any $\varepsilon > 0$ there exists a $\delta > 0$ such that $\mathscr{O}^+(\boldsymbol{\xi})$ is contained in the ε-neighborhood of $\boldsymbol{p}$ for any $\boldsymbol{\xi}$ in the δ-neighborhood of $\boldsymbol{p}$.
2. A fixed point is said to be unstable if it is not stable.
3. A fixed point $\boldsymbol{p}$ of f is said to be asymptotically stable if it is stable and there exists $\eta > 0$ such that

$$\lim_{n \to +\infty} f^n(\boldsymbol{\xi}) = \boldsymbol{p}$$

for any $\boldsymbol{\xi}$ with $|\boldsymbol{\xi} - \boldsymbol{p}| < \eta$.

Theorem 2.23 *Let $f : \mathbb{R}^n \to \mathbb{R}^n$ be a smooth map with a fixed point $\boldsymbol{p}$.*

1. *If all the eigenvalues of $f'(\boldsymbol{p})$ is inside the unit circle of the complex plane, then $\boldsymbol{p}$ is asymptotically stable.*
2. *If $f'(\boldsymbol{p})$ has at least one eigenvalue outside the unit circle of the complex plane, then $\boldsymbol{p}$ is unstable.*

Note that a periodic point $\boldsymbol{q}$ of f is a fixed point of f^m for some $m \in \mathbb{N}$. Applying the theorem above to the fixed point $\boldsymbol{q}$ of f^m, one can examine the stability of the periodic orbit $\mathscr{O}^+(\boldsymbol{q})$. A fixed point $\boldsymbol{p}$ is said to be hyperbolic if all the eigenvalues of $f'(\boldsymbol{p})$ is not on the unit circle. Similarly, a periodic point $\boldsymbol{q}$ with period m is said to be hyperbolic if it is a hyperbolic fixed point of f^m.

Bifurcations

See [24, 43] for details of local bifurcations of maps.

Theorem 2.24 *Let f be a smooth map $\mathbb{R} \times \mathbb{R} \ni (x, \alpha) \mapsto f(x, \alpha) \in \mathbb{R}$. Suppose that $f(\cdot, \alpha_0)$ has a fixed point x_0 for $\alpha_0 \in \mathbb{R}$ and $f_x(x_0, \alpha_0) = 1$. Assume that the following two conditions are satisfied:*

1. $f_{xx}(x_0, \alpha_0) \neq 0,$

2. $f_\alpha(x_0, \alpha_0) \neq 0$.

Then there is a smooth curve of fixed points in $\mathbb{R} \times \mathbb{R}$ *passing through* (x_0, α_0). *Depending on the signs of* $f_{xx}(x_0, \alpha_0)$ *and* $f_\alpha(x_0, \alpha_0)$, *there are no fixed point near* (x_0, α_0) *for* $\alpha < \alpha_0 (\alpha > \alpha_0)$, *and there are two fixed point near* (x_0, α_0) *for* $\alpha > \alpha_0 (\alpha < \alpha_0)$. *The two fixed points are both hyperbolic: one is stable and the other is unstable.*

The following theorem concerns the period doubling-bifurcation [24, Theorem 3.5.1]:

Theorem 2.25 *Let* $f_\mu : \mathbb{R} \to \mathbb{R}$ *be a one-parameter family of mappings such that* f_{μ_0} *has a fixed point* x_0 *with eigenvalue* -1. *Assume that the following two conditions hold at* (x_0, μ_0):

$$\left(\frac{\partial f}{\partial \mu} + 2\frac{\partial^2 f}{\partial x \partial \mu}\right) \neq 0,$$
$$\frac{1}{2}\left(\frac{\partial^2 f}{\partial x^2}\right)^2 + \frac{1}{3}\frac{\partial^3 f}{\partial x^3} \neq 0.$$

Then there is a smooth curve of fixed points of f_μ *passing through* (x_0, μ_0), *the stability of which changes at* (x_0, μ_0). *There is also a smooth curve* γ *passing through* (x_0, μ_0) *so that* $\gamma \setminus \{(x_0, \mu_0)\}$ *is a union of hyperbolic periodic orbits of period* 2. *The curve has quadratic tangency with the line* $\mathbb{R} \times \{\mu_0\}$ *at* (x_0, μ_0).

The following theorem concerns the Neimark–Sacker bifurcation [43, Theorems 4.5 and 4.6].

Theorem 2.26 *Let* f *be a smooth map* $\mathbb{R}^2 \times \mathbb{R} \mapsto f(\mathbf{x}, \alpha) \in \mathbb{R}^2$. *Suppose that* f *has a fixed point* $\mathbf{x} = \mathbf{0}$ *and* $D_x f(\mathbf{0}, 0)$ *has eigenvalues* $\mu(\alpha) = r(\alpha)e^{i\varphi(\alpha)}$ *and* $\overline{\mu(\alpha)}$ *with* $r(0) = 1, \varphi(0) = \theta_0$. *Assume the followings are satisfied:*

1. $r'(0) \neq 0$,
2. $e^{ik\theta_0} \neq 1$ *for* $k = 1, 2, 3, 4$.

Then there are smooth invertible coordinate and parameter changes transforming the system into

$$\begin{pmatrix} y_1 \\ y_2 \end{pmatrix} \mapsto (1 + \beta) \begin{pmatrix} \cos\theta(\beta) & -\sin\theta(\beta) \\ \sin\theta(\beta) & \cos\theta(\beta) \end{pmatrix} \begin{pmatrix} y_1 \\ y_2 \end{pmatrix}$$
$$+ (y_1^2 + y_2^2) \begin{pmatrix} \cos\theta(\beta) & -\sin\theta(\beta) \\ \sin\theta(\beta) & \cos\theta(\beta) \end{pmatrix} \begin{pmatrix} d(\beta) & -b(\beta) \\ b(\beta) & d(\beta) \end{pmatrix} \begin{pmatrix} y_1 \\ y_2 \end{pmatrix} + O(|\mathbf{y}|^4),$$

where $\mathbf{y} = (y_1, y_2)$, $\theta(0) = \theta_0$, *and* $d(\beta)$ *and* $b(\beta)$ *are real valued functions. Furthermore, if* $d(0) \neq 0$, *then there is a neighborhood of* $\mathbf{0}$ *in which a unique closed invariant curve bifurcates from* $\mathbf{0}$ *as* α *passes through zero.*

Continuation

Let $F : \mathbb{R}^n \times \mathbb{R} \to \mathbb{R}^n$ be a smooth map, and consider the set

$$S = \left\{(\boldsymbol{x}, \alpha) \in \mathbb{R}^n \times \mathbb{R} \mid F(\boldsymbol{x}, \alpha) = \boldsymbol{0}\right\}.$$

Suppose that we have a point $(\boldsymbol{x}_0, \alpha_0)$ on this set, that is, $F(\boldsymbol{x}_0, \alpha_0) = \boldsymbol{0}$. If the rank of the Jacobian matrix of F at $(\boldsymbol{x}_0, \alpha_0)$ is n, then the implicit function theorem implies that S is a curve in a sufficiently small neighborhood of $(\boldsymbol{x}_0, \alpha_0)$. Path-following methods enable us to compute the curve starting from the initial point $(\boldsymbol{x}_0, \alpha_0)$. Sophisticated numerical techniques for detecting bifurcation points on the curve and for switching the branch to be followed are available. See [43, Chap. 10] for details. AUTO [12] is one of the popular software in this subject, and it is employed in the present book.

Example 2.27 For a one-parameter family of smooth vector fields

$$\dot{\boldsymbol{x}} = f(\boldsymbol{x}, \alpha),$$

the set of equilibria is given by

$$\left\{(\boldsymbol{x}, \alpha) \in \mathbb{R}^n \times \mathbb{R} \mid f(\boldsymbol{x}, \alpha) = \boldsymbol{0}\right\}.$$

Example 2.28 For a one-parameter family of smooth maps

$$\boldsymbol{x} \mapsto f(\boldsymbol{x}, \alpha),$$

the set of fixed points is given by

$$\left\{(\boldsymbol{x}, \alpha) \in \mathbb{R}^n \times \mathbb{R} \mid \boldsymbol{x} - f(\boldsymbol{x}, \alpha) = \boldsymbol{0}\right\}.$$

In this case, $F(\boldsymbol{x}, \alpha) = \boldsymbol{x} - f(\boldsymbol{x}, \alpha)$.

Example 2.29 For a one-parameter family of smooth maps

$$\boldsymbol{x} \mapsto f(\boldsymbol{x}, \alpha),$$

the set of periodic points of period p is given by

$$\left\{(\boldsymbol{x}, \alpha) \in \mathbb{R}^n \times \mathbb{R} \mid \boldsymbol{x} - f^p(\boldsymbol{x}, \alpha) = \boldsymbol{0}\right\}.$$

In this case, $F(\boldsymbol{x}, \alpha) = \boldsymbol{x} - f^p(\boldsymbol{x}, \alpha)$.

Chapter 3
Mathematical Modeling, Simulation, and Analysis

Abstract The moving-boundary (MB) model is derived on the basis of physical and phenomenological considerations. A center manifold theory for the MB model is developed in this chapter. Consequently, under the assumption that the camphor disk is sufficiently slow, a system of ordinary differential equations is derived from the MB model in a mathematically rigorous manner. Ordinary differential equations that describe the interaction of two disks are also derived.

3.1 Mathematical Modeling

In this section a mathematical model for camphor motion is derived. An arbitrary shape is considered in [36, 58], while this book concentrates on a disk-shaped camphor scrapping. The model consists of a partial differential equation and an ordinary differential equation. Portions of this section are adapted from [9], used with permission of the American Institute of Mathematical Sciences, from Chen et al. [9]; permission conveyed through Copyright Clearance Center, Inc.

It is assumed that the camphor molecules dissolve into water, decreasing the surface tension, and the surface tension gradient drives the camphor motion. A camphor disk at time t is modeled by a disk $B(\boldsymbol{p}(t), r_0)$ centered at $\boldsymbol{x} = \boldsymbol{p}(t)$ with radius r_0. Let h_0 and ρ be the thickness and the density of the camphor disk, respectively. Two-dimensional motion of a camphor disk in a water tank is considered. The water tank is modeled by a domain Ω in $\mathbb{R}^2$. For simplicity, we assume $\Omega = \mathbb{R}^2$ hereafter. It is also assumed that the shape and size of the camphor disk does not change during the time interval of interest. Hence, r_0, h_0, and ρ are positive constants.

Consider a tri-intersection $\mathscr{C}$ among air, water (liquid), and the camphor disk (solid). Let θ be the contact angle between liquid and solid. In addition, let σ_{la}, σ_{ls} and σ_{sa} be surface tension at liquid-air, liquid-solid, and solid-air interfaces, respectively (see Fig. 3.1). Young's equality gives

$$\sigma_{ls} - \sigma_{sa} = -\sigma_{la} \cos\theta.$$

T. Miyaji et al., *A Billiard Problem in Nonlinear Dissipative Systems*,
Surveys and Tutorials in the Applied Mathematical Sciences 18,
https://doi.org/10.1007/978-981-95-7561-9_2_3

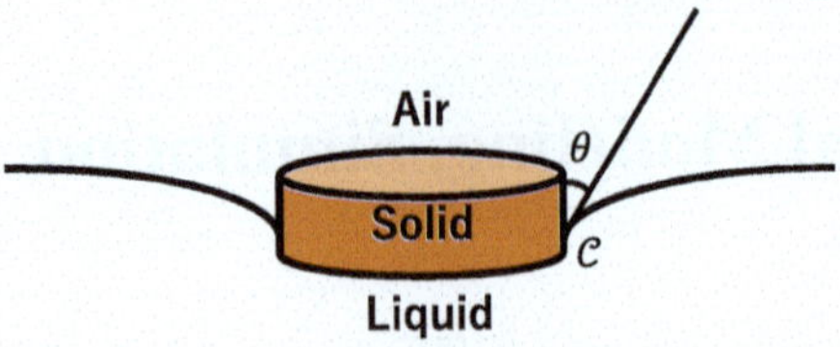

Fig. 3.1 A tri-interface among air, water (liquid), and the camphor disk (solid). Figure adapted from [9, Fig. 2], used with permission of the American Institute of Mathematical Sciences, from Chen et al. [9]; permission conveyed through Copyright Clearance Center, Inc.

The horizontal component of surface tension acting on the side of the camphor disk is

$$df_1 = \boldsymbol{n} ds \sigma_{la} \sin\theta,$$

where $\boldsymbol{n}$ is the outward unit normal vector on $\partial B(\boldsymbol{p}, r_0)$. The total contribution of the surface tension along $\mathscr{C}$ is given by

$$f_1 = \int_{\partial B(\boldsymbol{p},r_0)} \sigma_{la} \sin\theta \boldsymbol{n} ds.$$

Eliminating θ by Young's equality, we obtain

$$f_1 = \int_{\partial B(\boldsymbol{p},r_0)} \sigma_{la} \sqrt{1 - \left(\frac{\sigma_{ls} - \sigma_{sa}}{\sigma_{la}}\right)^2} \boldsymbol{n} ds.$$

As is well known that the inhomogeneous surface tension results in Marangoni flow, that is, a flow from lower to higher surface tension arises, we regard the contribution of the surface tension gradient over the entire disk as a driving force caused by the Marangoni flow. The total contribution can be calculated as

$$f_2 = \int_{B(\boldsymbol{p},r_0)} \nabla \sigma_{ls} dx.$$

By the divergence theorem, we obtain

$$f_2 = \int_{\partial B(\boldsymbol{p},r_0)} \sigma_{ls} \boldsymbol{n} ds.$$

Therefore the total contribution of surface tension to the camphor motion is

$$F = f_1 + f_2 = \int_{\partial B(\boldsymbol{p}, r_0)} \Gamma \boldsymbol{n} ds,$$

where

$$\Gamma = \sigma_{la} \sqrt{\frac{1 - (\sigma_{ls} - \sigma_{sa})^2}{\sigma_{la}^2}} + \sigma_{ls}.$$

We assume that Γ is a function of $c(t, \boldsymbol{x})$. According to the experiments by Ikura et al. [31], Γ is well approximated by

$$\Gamma(c) = \frac{\alpha^m \gamma_0}{\alpha^m + c^m} + \gamma_1$$

for some $\alpha > 0$, $\gamma_0 \geq 0$, $\gamma_1 \geq 0$ and $m \in \mathbb{N}$. In our simulation, we use $\alpha = 1/2$, $m = 1$, $\gamma_1 = 0$ for simplicity [9, 26].

It is assumed that the driving force of camphor motion is just the mechanical force described above and the camphor molecules at the water surface are transferred by diffusion. As mentioned above, the Marangoni flow is caused by a surface tension gradient and it can transfer camphor molecules at the surface. We concentrate on the case where the camphor disk moves very slowly, and we do not consider fluid mechanics here. It is also assumed that the camphor molecules are supplied from the camphor disk. Moreover, the camphor molecules are removed from the water surface because they sublimate at room temperature and dissolve slightly into water. Taking all of this into account, we describe the evolution of the camphor concentration at the water surface by

$$c_t = D\Delta c - \alpha c \chi_{\Omega \setminus B(\boldsymbol{p}, r_0)}(\boldsymbol{x}) + \tilde{F}(c)\chi_{B(\boldsymbol{p}, r_0)}(\boldsymbol{x}),$$

where D is the diffusion coefficient and α is the sublimation rate. The function χ is the characteristic function:

$$\chi_K(\boldsymbol{x}) = \begin{cases} 1 & (\boldsymbol{x} \in K), \\ 0 & (\boldsymbol{x} \notin K) \end{cases}$$

for $K \subseteq \Omega$. The function $\tilde{F}(c)$ represents the supply of camphor molecules. It can be chosen to be a constant function $\tilde{F}(c) = k > 0$. Let F be defined by

$$F(c) = \alpha c + \tilde{F}(c).$$

The equation is rewritten as

$$c_t = D\Delta c - \alpha c + F(c)\chi_{B(\boldsymbol{p}, r_0)}(\boldsymbol{x}).$$

If $\boldsymbol{x} \in B(\boldsymbol{p}, r_0)$, then $-\alpha c + F(c) = \tilde{F}(c)$, possibly $= k$. Therefore, the camphor molecules under the camphor disk do not sublimate.

Next, consider the motion of the camphor disk. Surface tension acts on the disk, while the viscosity of water gives a viscous force resisting the camphor motion. Assume that the viscous force is proportional to the velocity and the contact area of camphor disk. Namely, it is given by

$$\boldsymbol{F}_{\text{vis}} = -\mu\pi r_0^2 \dot{\boldsymbol{p}},$$

where μ is the viscosity coefficient. In this case, the equation of motion is

$$\begin{aligned}(\rho\pi r_0^2 h_0)\ddot{\boldsymbol{p}} &= \boldsymbol{F} + \boldsymbol{F}_{\text{vis}},\\ &= \int\limits_{\partial B(\boldsymbol{p},r_0)} \Gamma(c(t,\boldsymbol{x}))\boldsymbol{n}ds - \mu\pi r_0^2 \dot{\boldsymbol{p}}.\end{aligned}$$

In summary, our model is a system of a PDE and an ODE, given by

$$\begin{cases} c_t = D\Delta c - \alpha c + F(c)\chi_{B(\boldsymbol{p},r_0)}(\boldsymbol{x}), \\ (\rho\pi r_0^2 h_0)\ddot{\boldsymbol{p}} = \displaystyle\int\limits_{\partial B(\boldsymbol{p},r_0)} \Gamma(c(t,\boldsymbol{x}))\boldsymbol{n}ds - \mu\pi r_0^2 \dot{\boldsymbol{p}}. \end{cases} \tag{3.1}$$

If the camphor disk is very light and keeps constant velocity, inertia may be neglected. This would imply that

$$\int\limits_{\partial B(\boldsymbol{p},r_0)} \Gamma(c(t,\boldsymbol{x}))\boldsymbol{n}ds - \mu\pi r_0^2 \dot{\boldsymbol{p}} = 0.$$

Therefore, the equation for $\dot{\boldsymbol{p}}$ is determined by

$$\dot{\boldsymbol{p}} = \frac{1}{\mu\pi r_0^2} \int\limits_{\partial B(\boldsymbol{p},r_0)} \Gamma(c(t,\boldsymbol{x}))\boldsymbol{n}ds.$$

Thus, a simplified model is

$$\begin{cases} c_t = D\Delta c - \alpha c + F(c)\chi_{B(\boldsymbol{p},r_0)}(\boldsymbol{x}), \\ \dot{\boldsymbol{p}} = \beta \displaystyle\int\limits_{\partial B(\boldsymbol{p},r_0)} \Gamma(c(t,\boldsymbol{x}))\boldsymbol{n}ds, \end{cases} \tag{3.2}$$

where β is given by

$$\beta = \frac{1}{\mu\pi r_0^2}.$$

We call (3.2) the moving boundary (MB) model.

When we consider camphor motions in a bounded domain, we impose the Neumann boundary condition for equations involving c. As we shall see below, there exists a certain parameter domain in which the camphor disk does not collide with the boundary. In this book, we focus on the dynamics in such a parameter domain, and we do not consider the case where the camphor disk collides with the boundary.

The formulation above allows us to consider multiple camphor disks in a water tank. Suppose that there are m camphor disks with radii r_j, which are represented by $B(\boldsymbol{p}_j(t), r_j)$. The model equations are

$$\begin{cases} c_t = D\Delta c - \alpha c + F(c)\displaystyle\sum_{j=1}^{m} \chi_{B(\boldsymbol{p}_j, r_j)}(\boldsymbol{x}), \\ \dot{\boldsymbol{p}_j} = \beta_j \displaystyle\int_{\partial B(\boldsymbol{p}_j, r_j)} \Gamma(c(t,\boldsymbol{x}))\boldsymbol{n} ds, \quad (j = 1, \ldots, m), \end{cases} \tag{3.3}$$

where β_j is given by

$$\beta_j = \frac{1}{\mu\pi r_j^2}$$

for $j = 1, \ldots, m$. As mentioned above, we do not consider the collisions. The system (3.3) is called the moving boundary model for multiple disks.

3.2 Numerical Simulations

In this section, we describe a means of numerical simulation for the moving boundary model. First, let us consider a single camphor disk on a rectangle $\Omega = [-L, L] \times [-L, L]$ with periodic boundary conditions (PBC) for $L > 0$ instead of $\Omega = \mathbb{R}^2$. We express the function $c(t, \boldsymbol{x})$ by a Fourier series

$$c(t,\boldsymbol{x}) = \sum_{m=-\infty}^{\infty} \sum_{n=-\infty}^{\infty} c_{m,n}(t) \exp\left(\frac{m\pi \mathrm{i}x + n\pi \mathrm{i}y}{L}\right),$$

and consider the equations for the Fourier coefficients:

$$\dot{c}_{m,n} = \alpha_{m,n} c_{m,n} + g_{m,n}, \quad (m, n) \in \mathbb{Z} \times \mathbb{Z},$$

where $\alpha_{m,n}$ and g_n are given by

$$\alpha_{m,n} = \left[-\alpha - D\frac{(m^2+n^2)\pi^2}{L^2} \right],$$

$$g_{m,n} = \int_{\Omega} F(c(t,\boldsymbol{x}))\chi_{B(\boldsymbol{p},r_0)} \exp\left(-\frac{m\pi \mathrm{i}x + n\pi \mathrm{i}y}{L} \right) d\boldsymbol{x}.$$

We approximate $c(t, \boldsymbol{x})$ by truncating the Fourier series at $(m, n) = (\pm M, \pm N)$. Let us denote the resultant ODE by

$$\dot{U} = AU + G(U),$$

where U is a vector consisting of $c_{m,n}$ for $m = -M, \ldots, M$ and $n = -N, \ldots N$, and A is a diagonal matrix consisting of the $\alpha_{m,n}$ appearing above. Let $\Delta t > 0$.We take points $0 = t_0 < t_1 < t_2 < \ldots$ with $\Delta t = t_{j+1} - t_j > 0$. Let U_j be an approximate solution at time $t = t_j$. We apply Krogstad's exponential Runge–Kutta formula [42, p. 80] and [29, p. 228] to the initial-boundary value problem. Let $\varphi_0(z) = e^z$, and define φ_ℓ via the recurrence relation

$$\varphi_\ell(z) = \frac{\varphi_{\ell-1}(z) - \varphi_{\ell-1}(0)}{z}.$$

for $\ell = 1, 2, \ldots$. Following the notation of [29], the numerical scheme is given in the following way:

$$\begin{cases} k_1 = G(U_j), \\ k_2 = G\left(e^{\Delta t A/2}U_j + \Delta t \frac{1}{2}\varphi_{1,2}k_1\right), \\ k_3 = G\left(e^{\Delta t A/2}U_j + \Delta t \left\{\left(\frac{1}{2}\varphi_{1,3} - \varphi_{2,3}\right)k_1 + \varphi_{2,3}k_2\right\}\right), \\ k_4 = G\left(e^{\Delta t A}U_j + \Delta t \left\{(\varphi_{1,4} - 2\varphi_{2,4})k_1 + 2\varphi_{2,4}k_3\right\}\right), \\ U_{j+1} = e^{\Delta t A}U_j \\ \quad + \Delta t\{(\varphi_1 - 3\varphi_2 + 4\varphi_3)k_1 \\ \quad + (2\varphi_2 - 4\varphi_3)(k_2 + k_3) + (-\varphi_2 + 4\varphi_3)k_4\}, \end{cases}$$

where $\varphi_\ell = \varphi_\ell(\Delta t A)$, $\varphi_{1,2} = \varphi_{1,3} = \varphi_1((\Delta t/2)A)$, $\varphi_{2,3} = \varphi_2((\Delta t/2)A)$, $\varphi_{1,4} = \varphi_1$, and $\varphi_{2,4} = \varphi_2$. As $c(t, \cdot)$ is decomposed into Fourier modes, the matrix exponential $e^{\Delta t A}$ is diagonalized. Its diagonal entries are the exponentials $e^{-\Delta t \alpha_{m,n}}$. In addition, we apply the classical Runge–Kutta method to the equation for $\boldsymbol{p}$. We compute the φ-functions by contour integral, as proposed by Kassam and Trefethen [34]. The nonlinear term G is computed in the following way:

1. Transform U to c by using the Fast Fourier Transform (FFT).
2. Compute $g = F(c)\chi_{B(\boldsymbol{p},r_0)}$ for each grid point on $[-L, L] \times [-L, L]$.
3. Transform g to G by using the inverse FFT.

We use FFTW [20] for computing the FFT. The method above was used for simulations shown in Figs. 1.5, 1.6 and 1.7.

Next, let us consider simulations on a rectangle under the Neumann boundary condition, which corresponds to the case where the disk is confined in a rectangular region. In this case, we expand $c(t, \boldsymbol{x})$ via the eigenfunctions of the Laplacian on $[-kL, kL] \times [-L, L]$ with the Neumann boundary condition, that is,

$$c(t, \boldsymbol{x}) = \sum_{m=0}^{\infty} \sum_{n=0}^{\infty} c_{m,n}(t) \cos\left(\frac{m\pi x}{kL}\right) \cos\left(\frac{n\pi y}{L}\right).$$

In a similar manner to the PBC case, we truncated the series at $(m, n) = (M, N)$ and applied Krogstad's method. We solved the equations for $c_{m,n}(t)$ in the same way as above. We have implemented the solver so that the simulation stops if the disk intersects the boundary of domain. This method was used for Figs. 1.9, 1.10, 1.11, 1.12, 1.13, 1.14 and 1.15.

Figure 3.2 shows the relationship between a_0 and the speed of the disk on a square domain with the Neumann boundary condition. The maximum speed before reflection or collision is plotted for each a_0. The diagram is similar to the case of a flat torus. Compare it with Fig. 1.7. In a finite domain, it is expected that the rest state loses its stability by a Hopf bifurcation. If a_0 is far from the bifurcation point, and hence the disk moves quickly, then the disk may hit the boundary. Figure 3.3 shows snapshots just before a collision occurred in a case where $a_0 = 0.75$. One may compare this figure with Fig. 1.13. In fact, the disk hit the boundary for $a_0 \in (0.43, 1.0)$ in our numerical experiments. These threshold values of collision may depend on the domain size L, the initial condition, and boundary conditions.

Finally, we consider two disks on a rectangle $\Omega = [-L, L] \times [-L, L]$ with PBCs. In this case, we apply the above methods to (3.3) with $m = 2$. We solved the equations for U by Krogstad's method and the ODEs for $\boldsymbol{p}_1$ and $\boldsymbol{p}_2$ by the classical Runge–Kutta method. Figures 3.4 and 3.5 depict reflection of two disks for (3.3).

Obviously, the numerical method presented here relies heavily on the assumption that the domain is rectangular. To handle other domain shapes, a different numerical method is required.

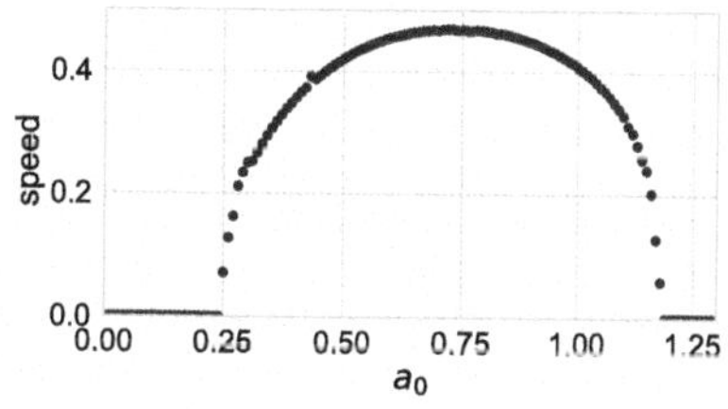

Fig. 3.2 Relationship between the purity a_0 and asymptotic speed in the MB model on a square domain

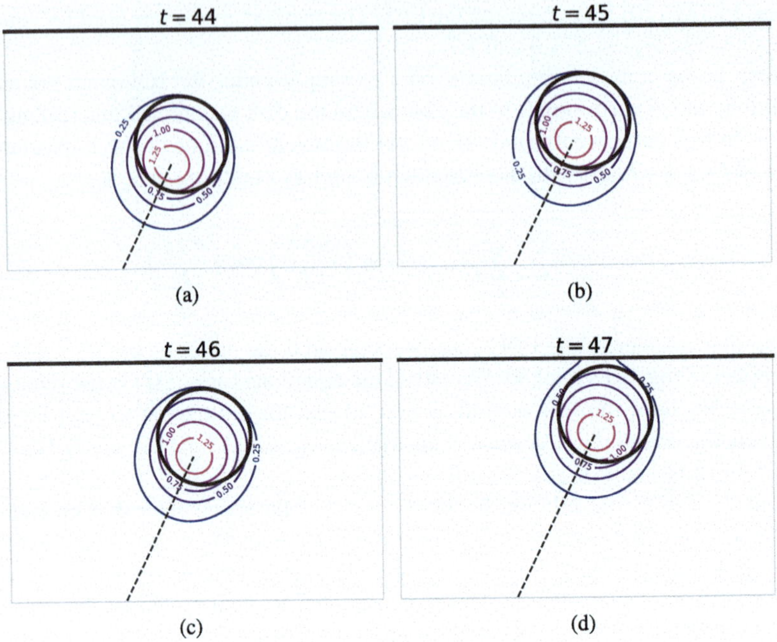

Fig. 3.3 Just before collision with the boundary ($a_0 = 0.75$). The black circle indicates the disk $B(t)$, the dashed line connected to the disk is the trajectory of the center, the colored curves are contour lines of $c(t, \boldsymbol{x})$, and the solid straight line is the boundary of the domain

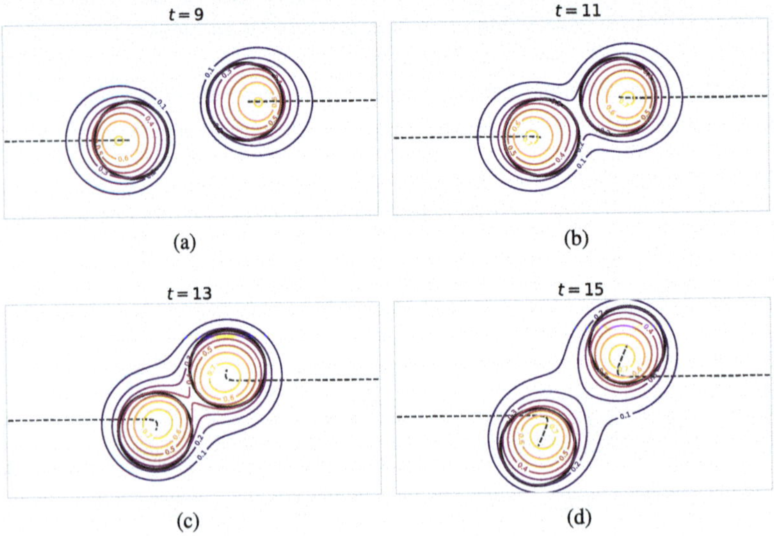

Fig. 3.4 Reflection of two disks

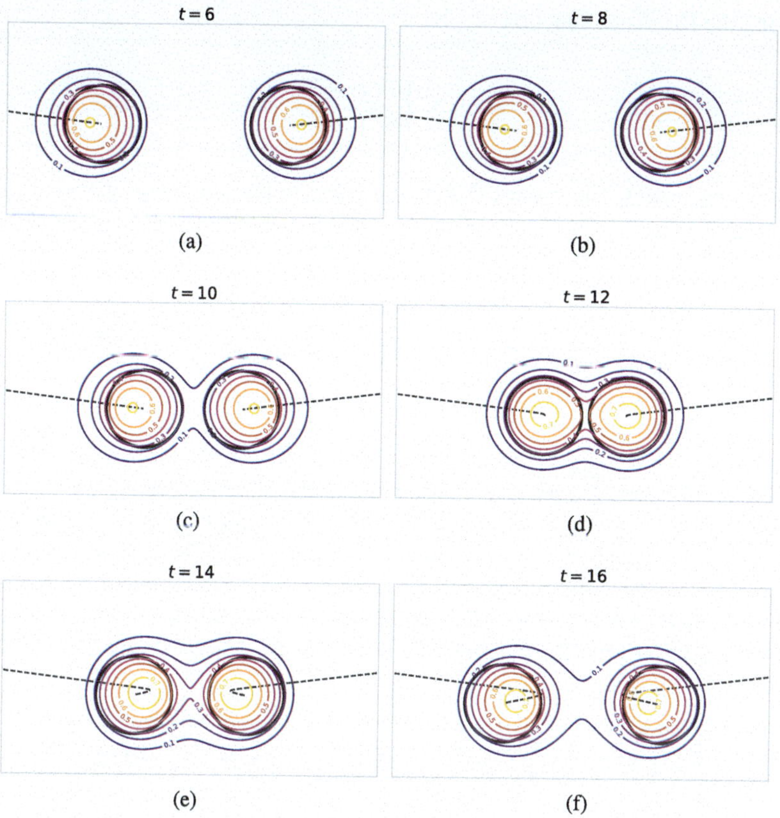

Fig. 3.5 Reflection of two disks which are at symmetric position with respect to the y-axis

3.3 Reduction Theory for a Single Disk

In this section, we mainly analyze the moving boundary model for a single disk (3.2) and theoretically derive the pitchfork bifurcation diagram of traveling spots as in Fig. 1.16. Portions of this section are adapted from [9], used with permission of the American Institute of Mathematical Sciences, from Chen et al. [9]; permission conveyed through Copyright Clearance Center, Inc. The moving boundary model for a single disk (3.2) can be expressed as

$$\begin{cases} c_t = D\Delta c - \alpha c + F(c)H(r_0 - |\boldsymbol{x} - \boldsymbol{p}|), & \boldsymbol{x} \in \mathbb{R}^2, t > \infty, \\ \dot{\boldsymbol{p}} = \beta \displaystyle\int_{\partial B(\boldsymbol{p}, r_0)} \Gamma(c(t, \boldsymbol{x}))\boldsymbol{n} ds \end{cases} \tag{3.4}$$

by using the Heaviside function

$$H(r) = \begin{cases} 1 & r > 0, \\ 0 & r < 0. \end{cases}$$

Let $\boldsymbol{y} = \boldsymbol{x} - \boldsymbol{p}(t)$. The system (3.4) is equivalent to the following system of equations:

$$\partial_t U = D\Delta U - \alpha U + \left\langle \beta \int_C \Gamma(U)\boldsymbol{n}ds, \nabla U \right\rangle + F(U)H(r_0 - |\boldsymbol{x}|), \tag{3.5}$$

$$\dot{\boldsymbol{p}} = \beta \int_C \Gamma(U)\boldsymbol{n}ds, \tag{3.6}$$

where $U(t, \boldsymbol{y}) = c(t, \boldsymbol{y} + \boldsymbol{p}(t))$, C is the circle centered at the origin with the radius r_0, and $\boldsymbol{n}$ is the unit outward normal vector of C. Note that (3.5) is a nonlocal equation but independent of $\boldsymbol{p}$. If we have a solution U of (3.5), then we obtain $\boldsymbol{p}(t)$ by integrating (3.6) with respect to t.

The rest state of a camphor disk is represented by a radially symmetric stationary solution of (3.5)–(3.6). Consider a function $S(r)$ on $[0, \infty)$ satisfying the following equations:

$$\begin{cases} D\left(S'' + \dfrac{1}{r}S'\right) - \alpha S + F(S)H(r_0 - r) = 0, \quad r \in [0, \infty) \\ S'(0) = 0, \\ \lim\limits_{r \to \infty} S(r) = 0. \end{cases} \tag{3.7}$$

For such $S(r)$, the pair $(U(t, \boldsymbol{x}), \boldsymbol{p}(t)) = (S(|\boldsymbol{x}|), \boldsymbol{0}) \in \mathbb{R} \times \mathbb{R}^2$ is a stationary solution of (3.5)–(3.6). Indeed, by the radial symmetry of U, the integrals in (3.5)–(3.6) vanish. The existence of S for (3.7) is assumed for a general function F, but we note that existence has been proven for some explicit examples of F, such as $F(c) = d_0$ and $F_0 - d_0 c$ for positive constants d_0 and F_0.

To analyze the linearization of (3.4) about $(U(t, \boldsymbol{x}), \boldsymbol{p}(t)) = (S(|\boldsymbol{x}|), \boldsymbol{0})$, let us consider (3.5). The linearized operator of the right hand side of (3.5) about $U = S(|\boldsymbol{x}|)$ is given by

$$L\phi = \{D\Delta - \alpha + F'(S)H(r_0 - |\boldsymbol{x}|)\}\phi + \left\langle \beta\Gamma'(S(r_0)) \int_C \phi\boldsymbol{n}ds, \nabla S \right\rangle. \tag{3.8}$$

Indeed, we have

$$\left\langle \int_C \Gamma(S+\varepsilon\phi)\boldsymbol{n}ds, \nabla(S+\varepsilon\phi)\right\rangle - \left\langle \int_C \Gamma(S)\nu ds, \nabla S\right\rangle$$
$$= \left\langle \int_C \Gamma(S+\varepsilon\phi)\boldsymbol{n}ds, \nabla(S+\varepsilon\phi)\right\rangle - \left\langle \int_C \Gamma(S)\boldsymbol{n}ds, \nabla(S+\varepsilon\phi)\right\rangle$$
$$+ \left\langle \int_C \Gamma(S)\boldsymbol{n}ds, \nabla(S+\varepsilon\phi)\right\rangle - \left\langle \int_C \Gamma(S)\boldsymbol{n}ds, \nabla S\right\rangle.$$

Dividing both sides by ε and taking the limit $\varepsilon \to 0$, we obtain

$$\left\langle \int_C \Gamma'(S)\phi\boldsymbol{n}ds, \nabla S\right\rangle + \left\langle \int_C \Gamma(S)\boldsymbol{n}ds, \nabla\phi\right\rangle.$$

Since $S(|\boldsymbol{x}|)$ is radially symmetric, the second term vanishes and $\Gamma'(S)$ is constant on C. Thus, we obtain (3.8) above.

First, we note that the essential spectrum of L is $\sigma_{\mathrm{ess}}(L) = (-\infty, -\alpha]$. Next, we consider the eigenvalues of L. We rewrite the eigenvalue equation $L\phi = \lambda\phi$ as

$$\begin{cases} \boldsymbol{V} = \beta\Gamma'(S(r_0)) \displaystyle\int_C \phi\boldsymbol{n}ds, \\ \left\{D\Delta + F'(S)H(r_0 - |\boldsymbol{x}|) - \alpha - \lambda\right\}\phi = -\langle \boldsymbol{V}, \nabla S\rangle. \end{cases} \tag{3.9}$$

Suppose that $\lambda \notin (-\infty, -\alpha]$. Then ϕ is uniquely determined for a given $\boldsymbol{V}$ because the left hand side of the second equation of (3.9) has an inverse for $\lambda \notin (-\infty, -\alpha]$ by $F' \leq 0$. We introduce the polar coordinates $\boldsymbol{x} = (r\cos\theta, r\sin\theta)$ and let $\boldsymbol{e}(\theta) = (\cos\theta, \sin\theta)$. Substituting the ansatz

$$\phi(r,\theta) = \langle \psi(r,\lambda)\boldsymbol{V}, \boldsymbol{e}(\theta)\rangle \tag{3.10}$$

into the second equation of (3.9), we notice that ψ is the unique solution of

$$\begin{cases} (L_R - \lambda)\psi = -S'(r), \\ \psi'(0,\lambda) = 0, \quad \psi(\infty,\lambda) = 0, \end{cases} \tag{3.11}$$

where L_R is defined by

$$L_R = D\left(\frac{d^2}{dr^2} + \frac{1}{r}\frac{d}{dr} - \frac{1}{r^2}\right) + F'(S(r))H(r_0 - r) - \alpha. \tag{3.12}$$

Define an inner product for $\psi_1, \psi_2 \in L^2(0, \infty)$ by

$$(\psi_1, \psi_2)_R = 2\pi \int_0^\infty r\psi_1(r)\overline{\psi_2}(r)dr.$$

Lemma 3.1 *Let b be a real number given by*

$$b = \frac{1}{\pi r_0 \beta |\Gamma'(S(r_0))|}.$$

A real number $\lambda \notin (-\infty, -\alpha]$ is an eigenvalue of L if and only if

$$-\psi(r_0, \lambda) = b. \tag{3.13}$$

Proof Substituting (3.10) into the first equation of (3.9), we have

$$V = \langle \beta\Gamma'(S(r_0)), \pi r_0 \psi(r_0, \lambda)V \rangle.$$

Here, we note that (3.10) leads to

$$\int_C \phi v ds = \int_C \psi(r, \lambda)\langle V, e(\theta)\rangle n ds = \pi r_0 \psi(r_0, \lambda)V.$$

Then, we have

$$1 = \beta\pi r_0 \Gamma'(S(r_0))\psi(r_0, \lambda)$$

and $\Gamma'(S(r_0)) < 0$ gives (3.13). □

Lemma 3.2 *All eigenvalues of L are real.*

Proof Note that L_R is negative definite. Differentiating (3.7) with respect to r, we obtain

$$L_R S' = F(S(r_0))\delta_{r_0},$$

where $\delta_{r_0}(r)$ is defined by $\delta_{r_0}(r) = \delta(r - r_0)$ for the Dirac δ function $\delta(r)$. Suppose that $\lambda \notin (-\infty, -\alpha]$ is an eigenvalue of L. Note that

$$(L_R\psi, S')_R = (\psi, L_R S')_R = 2\pi r_0 \psi(r_0, \lambda)F(S(r_0)).$$

Taking the inner product of (3.11) with $\lambda\psi$ and S', we obtain

$$\lambda\|\psi\|_R^2 = (S', \psi)_R + (L_R\psi, \psi)_R, \tag{3.14}$$

$$|\lambda|^2\|\psi\|_R^2 = \overline{\lambda}(S', \psi)_R + \overline{\lambda}(L_R\psi, \psi)_R, \tag{3.15}$$

$$\lambda(\psi, S')_R = \|S'\|_R^2 + 2\pi r_0 \psi(r_0, \lambda)F(S(r_0)). \tag{3.16}$$

By these equalities, we obtain

$$|\lambda|^2\|\psi\|_R^2 = \|S'\|_R^2 + 2\pi r_0\overline{\psi(r_0,\lambda)}F(S(r_0)) + \overline{\lambda}(L_R\psi,\psi)_R. \tag{3.17}$$

The previous lemma tells us that $\psi(r_0,\lambda) = -b \in \mathbb{R}$. In addition, $(L_R\psi,\psi)_R < 0$ because L_R is negative definite. Therefore, λ must be real. □

Lemma 3.3 *The spectrum of L is $\sigma(L) = (-\infty,-\alpha] \cup \{\lambda^*(b)\}$, where $\lambda^*(b)$ is an eigenvalue of L and is given by the solution of*

$$-\psi(r_0,\lambda) = b.$$

Moreover, $\lambda^*(b^*) = 0$ and $(\lambda^*)'(b^*) < 0$, where b^* is defined by

$$b^* = \frac{\|S'\|_R^2}{2\pi r_0 F(S(r_0))}.$$

In addition, $\lambda^*(b) < 0$ if and only if $b > b^*$.

Proof Since $L_R S' = F(S(r_0))\delta_{r_0}$, $S'(r)$ is negative for any $r > 0$. It implies that, for any $\lambda > -\alpha$ and $r \geq 0$, we have $\psi(r,\lambda) = -(L_R-\lambda)^{-1}S' < 0$. In addition, $|\psi(r_0,\lambda)|$ is decreasing with λ and $\psi(r_0,\lambda)$ tends to 0 as $\lambda \to \infty$. Therefore, for any $b > 0$ there exists a unique $\lambda = \lambda^*(b)$ such that $-\psi(r,\lambda) = b$ holds, where we define $\lambda^*(b) = -\alpha$ if $b > |\psi(r_0,-\alpha)|$. Since we have (3.16), $\psi(r,0) = -b^*$ for $\lambda = 0$. Moreover, since $\lambda \in \mathbb{R}$, it is easy to see by differentiating (3.17) with respect to λ and putting $\lambda = 0$ that

$$\lambda^*(b^*) = 0, \quad \frac{d}{db}\lambda^*(b^*) < 0,$$

and $\lambda^*(b) < 0$ if and only if $b > b^*$. □

Remark 3.4 We can rewrite the stability condition stated in Lemma 3.3 in the following way. Let β^* be defined by

$$\beta^* = \frac{1}{\pi r_0|\Gamma'(S(r_0))|b^*}.$$

Since $b = 1/(\pi r_0\beta|\Gamma'(S(r_0))|)$ by definition, $b > b^*$ holds if and only if

$$\beta < \beta^* = \frac{1}{\pi r_0|\Gamma'(S(r_0))|b^*} = \frac{2F(S(r_0))}{|\Gamma'(S(r_0))|\|S'\|_R^2}$$

holds. That is, the radially symmetric stationary solution is linearly stable if and only if $\beta < \beta^*$.

Remark 3.5 Suppose that $F(c) = \alpha c + a_0$, which is what has been used in the numerical simulations. For this function F, we can solve (3.7) explicitly and evaluate β_*. Let K_0 and K_1 be the zeroth and first modified Bessel function of the second kind, respectively. The solution of (3.7) is given by

$$S(r) = \begin{cases} S_0(r) & r \le r_0, \\ S_1(r) & r \ge r_0, \end{cases}$$

where S_0 and S_1 are solutions of

$$D\left(S_0'' + \frac{1}{r}S_0'\right) + a_0 = 0, \quad r \in [0, r_0],$$
$$D\left(S_1'' + \frac{1}{r}S_1'\right) - \alpha S_1 = 0, \quad r \in [r_0, \infty),$$

respectively. Taking the boundary conditions into account, we obtain

$$S_0(r) = a_0\left(c_0 - \frac{r^2}{4D}\right), \quad S_1(r) = a_0 c_1 K_0\left(\sqrt{\frac{\alpha}{D}}r\right).$$

Here, the constants c_0 and c_1 are

$$c_0 = \frac{2Dr_0\sqrt{\frac{\alpha}{D}}K_0\left(\sqrt{\frac{\alpha}{D}}r_0\right) + r_0^2\alpha K_1\left(\sqrt{\frac{\alpha}{D}}r_0\right)}{4D\alpha K_1\left(\sqrt{\frac{\alpha}{D}}r_0\right)},$$
$$c_1 = \frac{r_0}{2\alpha K_1\left(\sqrt{\frac{\alpha}{D}}r_0\right)}\sqrt{\frac{\alpha}{D}},$$

by which $S(r)$ is C^1 in $[0, \infty)$. See Fig. 3.6 for an example of $S(r)$ and its derivative. If $D = 0.2, \alpha = 1.0, r_0 = 1.0, \mu = 0.35$, then

$$\beta^* \approx 0.391551 + \frac{0.105548}{a_0} + 0.362134a_0, \quad \beta = \frac{1}{\mu\pi r_0^2} \approx 0.909457.$$

Solving $\beta^* = \beta$ with respect to a_0, we obtain $a_0 \approx 0.24635$ and 1.17986. On the other hand, solving $\beta^* = \beta$ with respect to r_0 in the case of $D = 0.2$, $\alpha = 1.0$, $a_0 = 1.0$, $\mu = 0.35$, we obtain $r_0 \approx 0.230997$ and 1.11121. These values are consistent with the simulations presented in Figs. 1.7, 1.8, and 3.2.

Based on the above discussion, we now consider the linearized operator, say $\boldsymbol{L}$, of (3.4) about the radially symmetric rest state $(c, \boldsymbol{p}) = (S(|\boldsymbol{x}|), \boldsymbol{0})$, which is given by

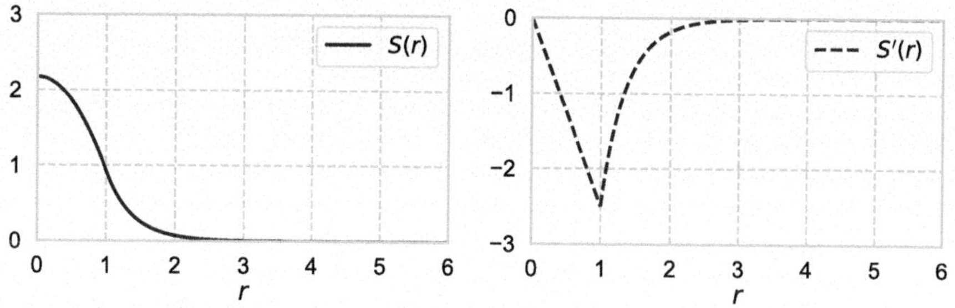

Fig. 3.6 $S(r)$ and $S'(r)$ in the case of $F(c) = \alpha c + a_0$ with $D = 0.2, \alpha = 1.0, r_0 = 1.0, a_0 = 0$

$$\boldsymbol{L}\boldsymbol{\Phi} = \boldsymbol{L}\begin{pmatrix}\phi \\ \boldsymbol{q}\end{pmatrix} = \begin{pmatrix} D\Delta - \alpha + F'(S)H(r_0 - r))\phi + F(S(r_0))\delta_C\langle \boldsymbol{e}(\theta), \boldsymbol{q}\rangle \\ \beta\Gamma'(S(r_0))[\pi r_0 S'(r_0)\boldsymbol{q} + \int_C \phi \boldsymbol{n} ds] \end{pmatrix}$$

for $\boldsymbol{\Phi} = (\phi, \boldsymbol{q}) \in H^2(\mathbb{R}^2) \times \mathbb{R}^2$.

Let $\boldsymbol{\Phi}_1$ and $\boldsymbol{\Phi}_2$ be given by

$$\boldsymbol{\Phi}_1(r, \theta) = \begin{pmatrix} S'(r)\cos\theta \\ -\boldsymbol{e}_1 \end{pmatrix}, \quad \boldsymbol{\Phi}_2(r, \theta) = \begin{pmatrix} S'(r)\sin\theta \\ -\boldsymbol{e}_2 \end{pmatrix},$$

where $\boldsymbol{e}_1 = (1, 0)$ and $\boldsymbol{e}_2 = (0, 1)$. First, we note that $\boldsymbol{\Phi}_1$ and $\boldsymbol{\Phi}_2$ satisfy $\boldsymbol{L}\boldsymbol{\Phi}_1 = 0$ and $\boldsymbol{L}\boldsymbol{\Phi}_2 = 0$, respectively. Thus, 0 is always an eigenvalue of $\boldsymbol{L}$ due to the translation free mode of the system (3.4).

Next, we consider the eigenvalue problem $\boldsymbol{L}\boldsymbol{\Phi} = \lambda\boldsymbol{\Phi}$, which can be written

$$\boldsymbol{L}\begin{pmatrix}\phi \\ \boldsymbol{q}\end{pmatrix} = \begin{pmatrix} \{L_R\psi + F(S(r_0))\delta(r_0 - r)\}\langle \boldsymbol{e}(\theta), \boldsymbol{q}\rangle \\ \beta\Gamma'(S(r_0))[\pi r_0 S'(r_0)\boldsymbol{q} + \pi r_0 \psi(r_0)\boldsymbol{q}] \end{pmatrix} = \lambda \begin{pmatrix} \psi\langle \boldsymbol{e}(\theta), \boldsymbol{q}\rangle \\ \boldsymbol{q} \end{pmatrix} \tag{3.18}$$

for $\boldsymbol{q} \in \mathbb{R}^2$ and $\phi(r, \theta) = \psi(r)\langle \boldsymbol{e}(\theta), \boldsymbol{q}\rangle$.

Let $\widetilde{\psi}(\cdot; \lambda) = -F(S(r_0))(L_R - \lambda)^{-1}\delta_{r_0}$. The first equation of (3.18) leads to $\psi = \widetilde{\psi}(\cdot; \lambda)$ and the first equation together with $\widetilde{\psi}$ gives $S'(r_0) + \widetilde{\psi}(r_0) = -\lambda b$ with b as given in Lemma 3.3. That is, we see that λ is an eigenvalue of $\boldsymbol{L}$ if and only if

$$\begin{cases} (L_R - \lambda)\widetilde{\psi}(r) = -F(S(r_0))\delta(r - r_0), \\ \lambda b + S'(r_0) + \widetilde{\psi}(r_0) = 0 \end{cases} \tag{3.19}$$

holds.

The inner product of (3.19) with $\widetilde{\psi}$ implies

$$-2\pi r_0 F(S(r_0))\overline{\widetilde{\psi}(r_0)} = (L_R\widetilde{\psi}, \widetilde{\psi})_R - \lambda\|\widetilde{\psi}\|_R^2 \tag{3.20}$$

and the inner product with $S'(r)$

$$\lambda(\widetilde{\psi}, S')_R = (L_R\widetilde{\psi}, S')_R + 2\pi r_0 F(S(r_0))S'(r_0).$$

Since $L_R S' = F(S(r_0))\delta_{r_0}$ holds by Lemma 3.2, we have

$$(L_R\widetilde{\psi}, S')_R = (\widetilde{\psi}, L_R S')_R = F(S(r_0)) \cdot 2\pi r_0 \widetilde{\psi}(r_0)$$

and hence

$$\lambda(\widetilde{\psi}, S')_R = 2\pi r_0 F(S(r_0))\{\widetilde{\psi}(r_0) + S'(r_0)\}. \tag{3.21}$$

(3.21) and the second equation of (3.19) lead to

$$\lambda\{(\widetilde{\psi}, S')_R + 2\pi r_0 F(S(r_0))b\} = 0. \tag{3.22}$$

(3.22) implies that non-zero λ satisfies

$$(\widetilde{\psi}, S')_R + 2\pi r_0 F(S(r_0))b = 0, \tag{3.23}$$

in particular, $(\widetilde{\psi}, S')_R \in \mathbb{R}$ holds and also $\lambda \in \mathbb{R}$.

Since $S'(r) < 0$, $\widetilde{\psi}(r; \lambda) > 0$ and $\widetilde{\psi}(r; \lambda)$ decreases monotonically with respect to λ for $\lambda > -\alpha$, as stated in the proof of Lemma 3.3, we see that $-(\widetilde{\psi}(\cdot; \lambda), S')_R$ is monotone decreasing with respect to λ.

We define $\lambda = \widetilde{\lambda^*}(b)$ by the solution of (3.23). Here we put $\widetilde{\lambda^*}(b) = -\alpha$ when $-(\widetilde{\psi}(\cdot; -\alpha), S')_R < 2\pi r_0 F(S(r_0))b$. We note that $\widetilde{\lambda^*}(b^*) = 0$ holds due to $\widetilde{\psi}(r; 0) = -S'$ and (3.23).

Quite similarly, it is shown that $\widetilde{\lambda^*}(b) < 0$ for $b > b^*$ and $\widetilde{\lambda^*}(b) > 0$ for $b < b^*$.

Now we consider eigenfunctions associated with the 0 eigenvalue of $\boldsymbol{L}$. Note that

$$\ker \boldsymbol{L} = \mathrm{span}\{\boldsymbol{\Phi}_1, \ \boldsymbol{\Phi}_2\}$$

holds for any b because $\boldsymbol{L\Phi} = 0$ for $\boldsymbol{\Phi} = (\phi, \boldsymbol{q})$ implies $\phi(r, \theta) = \widetilde{\psi}(r; 0)\langle \boldsymbol{e}(\theta), \boldsymbol{q}\rangle$ by (3.19).

Define ϕ_0, Ψ_1, and Ψ_2 by $\phi_0 = -(L_R)^{-1}S'$ and

$$\Psi_1(r, \theta) = \begin{pmatrix} \phi_0(r)\cos\theta \\ \mathbf{0} \end{pmatrix}, \quad \Psi_2(r, \theta) = \begin{pmatrix} \phi_0(r)\sin\theta \\ \mathbf{0} \end{pmatrix}.$$

Then we have the lemma:

Lemma 3.6 *For $b = b^*$ ($\beta = \beta^*$), $\boldsymbol{\Phi}_1$, $\boldsymbol{\Phi}_2$, Ψ_1 and Ψ_2 satisfy*

$$\mathbf{L}_0\boldsymbol{\Phi}_1 = \mathbf{L}_0\boldsymbol{\Phi}_2 = 0, \quad \mathbf{L}_0\Psi_1 = -\boldsymbol{\Phi}_1 \quad \mathbf{L}_0\Psi_2 = -\boldsymbol{\Phi}_2,$$

where $\mathbf{L}_0$ is the operator $\mathbf{L}$ at $\beta = \beta^*$.

Proof Ψ_1 satisfies

$$\boldsymbol{L}_0\Psi_1 = \boldsymbol{L}_0\begin{pmatrix}\phi_0(r)\cos\theta\\ \boldsymbol{0}\end{pmatrix} = \begin{pmatrix}(L_R\phi_0)\cos\theta\\ \beta^*\Gamma'(S(r_0))\phi_0(r_0)\int_C\cos\theta\boldsymbol{n}ds\end{pmatrix}$$

$$= \begin{pmatrix}-S'\cos\theta\\ \beta^*\Gamma'(S(r_0))\phi_0(r_0)\pi r_0\boldsymbol{e}_1\end{pmatrix}.$$

Since b^* satisfies $\phi_0(r_0) = \psi(r_0; 0) = -|\psi(r_0; 0)| = -b^*$ (see Lemma 3.3), we have

$$\beta^*\Gamma'(S(r_0))\phi_0(r_0)\pi r_0 = -\beta^*\Gamma'(S(r_0))b^*\pi r_0.$$

Then the definition of β^* and $\Gamma' \leq 0$ lead to

$$-\beta^*\Gamma'(S(r_0))b^*\pi r_0 = \beta^*|\Gamma'(S(r_0))|b^*\pi r_0 = 1,$$

which means

$$\boldsymbol{L}_0\Psi_1 = \begin{pmatrix}-S'\cos\theta\\ \boldsymbol{e}_1\end{pmatrix} = -\Phi_1.$$

Other cases are analogous. □

Finally, we consider the adjoint operator $\boldsymbol{L}^*$ of $\boldsymbol{L}$, in particular, $\boldsymbol{L}_0^*$, the operator $\boldsymbol{L}^*$ at $b = b^*$ ($\beta = \beta^*$), which is given by

$$\boldsymbol{L}_0^*\Phi = \boldsymbol{L}_0^*\begin{pmatrix}\phi\\ \boldsymbol{q}\end{pmatrix} = \begin{pmatrix}(d\Delta - \alpha + F'(S)H)\phi + \beta^*\Gamma'(S(r_0))\delta_C\langle\boldsymbol{e}(\theta), \boldsymbol{q}\rangle\\ \beta^*\Gamma'(S(r_0))\pi r_0 S'(r_0)\boldsymbol{q} + F(S(r_0))\int_C\phi\boldsymbol{n}ds\end{pmatrix}$$

for $\Phi = (\phi, \boldsymbol{q}) \in H^2(\mathbb{R}^2) \times \mathbb{R}^2$. Then the following lemma holds:

Lemma 3.7 *Let $b = b^*$ ($\beta = \beta^*$) and*

$$\Phi_1^*(r, \theta) = \begin{pmatrix}S'(r)\cos\theta\\ -\frac{F(S(r_0))}{\beta^*\Gamma'(S(r_0))}\boldsymbol{e}_1\end{pmatrix}, \quad \Phi_2^*(r, \theta) = \begin{pmatrix}S'(r)\sin\theta\\ -\frac{F(S(r_0))}{\beta^*\Gamma'(S(r_0))}\boldsymbol{e}_2\end{pmatrix}$$

and

$$\Psi_1^*(r, \theta) = \begin{pmatrix}(-a_0S'(r) + \phi_0(r))\cos\theta\\ \frac{a_0F(S(r_0))}{\beta^*\Gamma'(S(r_0))}\boldsymbol{e}_1\end{pmatrix}, \quad \Psi_2^*(r, \theta) = \begin{pmatrix}(-a_0S'(r) + \phi_0(r))\sin\theta\\ \frac{a_0F(S(r_0))}{\beta^*\Gamma'(S(r_0))}\boldsymbol{e}_2\end{pmatrix},$$

where $a_0 = \|\phi_0\|_R^2/(S', \phi_0)_R$. Then $\boldsymbol{L}_0^*\Phi_j^* = 0$ and $\boldsymbol{L}_0^*\Psi_j^* = -\Phi_j^*$ hold with the normalization

$$(\Phi_j, \Phi_j^*)_{L^2} = (\Psi_j, \Psi_j^*)_{L^2} = 0,$$

and

$$(\Psi_j, \Phi_j^*)_{L^2} = (\Phi_j, \Psi_j^*)_{L^2} = \frac{1}{2}\langle S', \phi_0 \rangle_R > 0,$$

together with

$$(\Phi_j, \Phi_k^*)_{L^2} = (\Phi_j, \Psi_k^*)_{L^2} = (\Psi_j, \Phi_k^*)_{L^2} = (\Psi_j, \Psi_k^*)_{L^2} = 0$$

for $j \neq k$.

This Lemma is easily checked by substituting functions into $\boldsymbol{L}_0^*$, noting

$$\frac{1}{2}\|S'\|_R^2 + \frac{F(S(r_0))}{\beta^* \Gamma'(S(r_0))} = 0$$

from the definition of β^* in Remark 3.4. We omit the proof.

Finally, we consider the eigenvalue λ^* in a neighborhood of $b = b^*$. Unfortunately, b is not easy to deal with as a bifurcation parameter. Hence we use β as the bifurcation parameter hereafter, as stated in Remark 3.4, and rewrite $\lambda^*(b)$ as $\lambda^*(\beta)$. Then we note that for $\beta^* = 1/(\pi r_0|\Gamma'(S(r_0))|b^*)$,

$$\lambda^*(\beta) \begin{cases} < 0 & (\beta < \beta^*), \\ = 0 & (\beta = \beta^*), \\ > 0 & (\beta > \beta^*), \end{cases}$$

hold, and also Lemmas 3.6 and 3.7 at $\beta = \beta^*$. These properties means that the assumptions (H1), (H2) and (H3) in [16] hold for the system (3.4). Therefore, the bifurcation of a traveling spot solution and its interaction can be analyzed in a general framework by applying the results in [16]. Hereafter, we only give the outline.

Define the map $\tau(\boldsymbol{p})$ by the translation

$$\tau(\boldsymbol{p})\begin{pmatrix} c(\boldsymbol{x}) \\ \boldsymbol{q} \end{pmatrix} = \begin{pmatrix} c(\boldsymbol{x} - \boldsymbol{p}) \\ \boldsymbol{q} \end{pmatrix},$$

and define Φ_j and Ψ_j by

$$\Phi_j(\boldsymbol{p}) = \tau(\boldsymbol{p})\Phi_j, \quad \Psi_j(\boldsymbol{p}) = \tau(\boldsymbol{p})\Psi_j$$

for $j = 1, 2$. Lut us introduce the following notation:

$$\boldsymbol{S}(\boldsymbol{x}) = \begin{pmatrix} S(|\boldsymbol{x}|) \\ \boldsymbol{0} \end{pmatrix}, \quad \boldsymbol{S}(\boldsymbol{x}; \boldsymbol{p}) = \begin{pmatrix} S(|\boldsymbol{x} - \boldsymbol{p}|) \\ \boldsymbol{p} \end{pmatrix},$$

$$\mathcal{M} = \left\{\boldsymbol{S}(\cdot; \boldsymbol{p}) \mid \boldsymbol{p} \in \mathbb{R}^2\right\},$$

$$\boldsymbol{U}^*(\boldsymbol{p}, \boldsymbol{v}) = \boldsymbol{S}(\boldsymbol{p}) + v\Psi_1(\boldsymbol{p}) + w\Psi_2(\boldsymbol{p}) \quad (\boldsymbol{v} = (v, w)).$$

We define the subspace spanned by Φ_j and Ψ_j:

$$E = \text{span}\{\Phi_1,\ \Phi_2,\ \Psi_1,\ \Psi_2\},$$
$$E(\boldsymbol{p}) = \tau(\boldsymbol{p})E = \text{span}\{\Phi_1(\boldsymbol{p}),\ \Phi_2(\boldsymbol{p}),\ \Psi_1(\boldsymbol{p}),\ \Psi_2(\boldsymbol{p})\}.$$

We also define the orthogonal space

$$E^{\perp} = \left\{\boldsymbol{V} \in H^2(\mathbb{R}^2) \times \mathbb{R}^2 \mid (\boldsymbol{V}, \Phi_j^*)_{L^2} = (\boldsymbol{V}, \Psi_j^*)_{L^2} = 0\ (j = 1, 2)\right\}$$

and

$$\begin{aligned} E^{\perp}(\boldsymbol{p}) &= \tau(\boldsymbol{p})E^{\perp} \\ &= \left\{\boldsymbol{V} \in H^2(\mathbb{R}^2) \times \mathbb{R}^2 \mid (\boldsymbol{V}, \Phi_j^*(\boldsymbol{p}))_{L^2} = (\boldsymbol{V}, \Psi_j^*(\boldsymbol{p}))_{L^2} = 0\ (j = 1, 2)\right\}. \end{aligned}$$

Then, $\boldsymbol{U} = (u, \widehat{\boldsymbol{p}})$ in the neighborhood of $\mathscr{M}$ is uniquely represented by

$$\boldsymbol{U} = \boldsymbol{U}^*(\boldsymbol{p}, \boldsymbol{v}) + \boldsymbol{V}$$

for $\boldsymbol{p} \in \mathbb{R}^2$ and $\boldsymbol{V} \in E^{\perp}(\boldsymbol{p})$ with $\boldsymbol{V}(\boldsymbol{x}) = (v(\boldsymbol{x}), \boldsymbol{q})$. By using this decomposition, we write the solution

$$\boldsymbol{U}(t, \boldsymbol{x}) = \boldsymbol{U}^*(\boldsymbol{x}; \boldsymbol{p}(t), \boldsymbol{v}(t)) + \boldsymbol{V}(t, \boldsymbol{x})$$

with $\boldsymbol{V}(t) \in E^{\perp}(\boldsymbol{p}(t))$. For convenience, we also write it as

$$\boldsymbol{U}(t, \boldsymbol{x}) = \boldsymbol{S}(\boldsymbol{x}; \boldsymbol{p}(t)) + \boldsymbol{W}(t, \boldsymbol{x})$$

using

$$\boldsymbol{W}(t, \boldsymbol{x}) = v(t)\Psi_1(\boldsymbol{p}(t)) + w(t)\Psi_2(\boldsymbol{p}(t)) + \boldsymbol{V}(t, \boldsymbol{x}).$$

By $\beta = \beta^* + \eta$, (3.4) becomes

$$\begin{cases} c_t(t, \boldsymbol{x}) = D\Delta c - \alpha c + F(c)H(r_0 - |\boldsymbol{x} - \boldsymbol{p}|), \\ \dfrac{d}{dt}\boldsymbol{p} = \beta^* \displaystyle\int_C \Gamma(c(\boldsymbol{p} + \boldsymbol{y}))\boldsymbol{n} ds + \eta \int_C \Gamma(c(\boldsymbol{p} + \boldsymbol{y}))\boldsymbol{n} ds. \end{cases} \tag{3.24}$$

(3.24) is compactly represented by

$$\boldsymbol{U}_t = \boldsymbol{A}_0(\boldsymbol{U}) + \eta \boldsymbol{G}(\boldsymbol{U}) \tag{3.25}$$

for $\boldsymbol{U} = (c, \boldsymbol{p})$, where

$$A_0(U) = \begin{pmatrix} D\Delta c - \alpha c + F(c)H(r_0 - |x - p|) \\ \beta^* \int_C \Gamma(c(p+y))nds \end{pmatrix},$$

$$G(U) = \begin{pmatrix} 0 \\ \int_C \Gamma(c(p+y))nds \end{pmatrix}.$$

$A_0(S(p)) = \mathbf{0}$ holds for $p \in \mathbb{R}^2$ and the linearized operator $L_0(p) = A_0'(S(p))$ satisfies

$$L_0(p)\begin{pmatrix}\phi \\ q\end{pmatrix} = \tau(p)L_0\begin{pmatrix}\phi \\ q\end{pmatrix}$$
$$= \begin{pmatrix} D\Delta - \alpha + F'(S)H(r_0 - r)\phi + F(S(r_0))\delta_C\langle e(\theta), q\rangle \\ \beta^*\Gamma'(S(r_0))[\pi r_0 S'(r_0)q + \int_C \phi n ds] \end{pmatrix}$$

for $x - p = re(\theta)$, where we note that L_0 is the operator L at $\beta = \beta^*$ corresponding to $L_0(\mathbf{0})$. Then, the assumptions (H1) to (H3) in [16] are all satisfied for (3.24). That is, the following three assumptions are satisfied by Lemmas 3.3, 3.6 and 3.7.

(H1) $A_0(S(p)) = \mathbf{0}$ holds for p.

(H2) The spectral set $\Sigma(L_0)$ of L_0 consists of the two sets

$$\Sigma_0 = \{0\} \quad \text{and} \quad \Sigma_1 \subset \{z \in \mathbb{C} \mid \mathrm{Re}(z) < -\gamma_0\}$$

for a positive constant γ_0. The generalized eigenspace corresponding to Σ_0 is given by $E = \mathrm{span}\{\Phi_1, \Phi_2, \Psi_1, \Psi_2\}$, satisfying $L_0\Phi_j = 0$ and $L_0\Psi_j = -\Phi_j$ for $j = 1, 2$.

(H3) There are radially symmetric functions $\Psi_0(r)$, $\Phi_0^*(r)$ and $\Psi_0^*(r)$ such that the eigenfunctions Ψ_j, Φ_j^* and Ψ_j^* are given by their derivatives. That is, $\Psi_j = \partial_{p_j}\{\tau(p)\Psi_0\}|_{p=\mathbf{0}}$ for $\Psi_0(r) = (\int \phi_0 dr, \mathbf{0})$ and so on, using the operator $\tau(p)$ defined by

$$\tau(p)\begin{pmatrix} c(x) \\ q\end{pmatrix} = \begin{pmatrix} c(x-p) \\ q+p \end{pmatrix}.$$

Then we can apply the results of [16] to Eq. (3.25). As a consequence, the following results corresponding to Theorem 2.2 in [16] hold for general types of equations like (3.25) when (H1), (H2) and (H3) hold:

Theorem 3.8 *([16]) The solution $U(t, x)$ of* (3.24) *stays close to $U^*(p(t), v(t))$ for $t > 0$ as long as $|v(t)| < v^*$ and $|\eta| < \eta^*$ for constants $v^* > 0$ and $\eta^* > 0$ when the initial data $U(0, x)$ is sufficiently close to $S(x; p_0)$ for $p_0 \in \mathbb{R}^2$. $p(t)$ and $v(t)$ satisfy*

$$\begin{cases} \dot{p} = v + O(|v|^3 + |\eta|^{\frac{3}{2}}), \\ \dot{v} = -\nabla_v W + O(|v|^4 + |\eta|^2), \end{cases} \tag{3.26}$$

where $W = W(v) = \frac{1}{4}M_1|v|^4 + \frac{1}{2}M_2\eta|v|^2$. *The constants are given by*

$$(\Psi_1, \Phi_1^*)_{L^2} M_1 = -\left\{\frac{1}{6}(A_0'''(S(r))\Psi_1^3, \Phi_1^*)_{L^2} + (A_0''(S(r))\Psi_1 V_1, \Phi_1^*)_{L^2}\right.$$
$$\left. -(\partial_{p_1}\{\tau(p)V_1\}|_{p=0}, \Phi_1^*)_{L^2}\right\},$$
$$(\Psi_1, \Phi_1^*)_{L^2} M_2 = -\left\{(G'(S)\Psi_1, \Phi_1^*)_{L^2} + (A_0''(S(r))\Psi_1 V_2, \Phi_1^*)_{L^2}\right.$$
$$\left. -(\partial_{p_1}\{\tau(p)V_2\}|_{p=0}, \Phi_1^*)_{L^2}\right\},$$

where the functions V_1 *and* V_2 *in* $E^\perp$ *are those satisfying*

$$-L_0 V_1 = \frac{1}{2}A_0''(S)\Psi_1^2 - \partial_{p_1}\{\tau(p)\Psi_1\}|_{p=0},$$
$$-L_0 V_2 = G(S).$$

Lemma 3.9 M_1 *and* M_2 *for the equation (3.24) are given by*

$$M_1 = -\frac{2\pi}{(S', \phi_0)_R}\left(\frac{1}{8}\left\{\int_0^{r_0} rF'''(S(r))\phi_0^3(r)S'(r)dr - \frac{\Gamma'''(S(r_0))}{\Gamma'(S(r_0))}\phi_0^3(r_0)F(S(r_0))\right\}\right.$$
$$+\left\{\frac{1}{4}\int_0^{r_0} rF''(S(r))\phi_0(r)(Z_0(r) + \frac{1}{2}Z_2(r))S'(r)dr\right.$$
$$\left.-\frac{\Gamma''(S(r_0))F(S(r_0))\phi_0(r_0)}{4\Gamma'(S(r_0))}(Z_0(r_0) + \frac{1}{2}Z_2(r_0))\right\}$$
$$\left.+\frac{1}{2}\int_0^{r_0}(2rZ_0'(r) + rZ_2'(r) + 2Z_2(r))S'(r)dr\right),$$
$$M_2 = \frac{2\phi_0(r_0)F(S(r_0))}{\beta^*(S', \phi_0)_R},$$

where $Z_0(r)$ *and* $Z_2(r)$ *are the solutions of*

$$-L_R^0 Z_0 = \frac{1}{2}\left(\frac{1}{2}F''(S(r))H(r_0 - r)\phi_0^2 + \phi_0' + \frac{1}{r}\phi_0(r)\right), \tag{3.27}$$
$$-L_R^2 Z_2 = \frac{1}{2}\left(\frac{1}{2}F''(S(r))H(r_0 - r)\phi_0^2 + \phi_0' - \frac{1}{r}\phi_0(r)\right), \tag{3.28}$$

respectively, and

$$L_R^N = D\left(\frac{d^2}{dr^2} + \frac{1}{r}\frac{d}{dr} - \frac{N^2}{r^2}\right) + F'(S(r))H(r_0 - r) - \alpha.$$

Proof The proof is a direct result of substituting results of Lemmas 3.3, 3.6 and 3.7 and applying Theorem 3.8. We simply show that the constants M_1 and M_2 are given by the above equations.

First, we shall consider M_1. We calculate

$$(\boldsymbol{A}_0'''(\boldsymbol{S})\Psi_1^3, \Phi_1^*)_{L^2} = \left(\begin{pmatrix} \cos^3\theta F'''(S(r))H(r_0 - r)\phi_0^3(r) \\ \beta^*\Gamma'''(S(r_0)) \int_C \cos^3\theta\phi_0(r)\boldsymbol{n}ds \end{pmatrix}, \begin{pmatrix} S'\cos\theta \\ -\frac{F(S(r_0))}{\beta^*\Gamma'(S(r_0))}\boldsymbol{e}_1 \end{pmatrix} \right)_{L^2}$$
$$= \frac{3}{4}\pi M_1',$$

where

$$M_1' = \int_0^{r_0} rF'''(S(r))\phi_0^3(r)S'(r)dr - \frac{\Gamma'''(S(r_0))}{\Gamma'(S(r_0))}\phi_0^3(r_0)F(S(r_0)).$$

Thus we obtain

$$\frac{1}{6}(\boldsymbol{A}_0'''(\boldsymbol{S})\Psi_1^3, \Phi_1^*)_{L^2} = \frac{1}{8}\pi M_1'. \tag{3.29}$$

Next, consider the term

$$(\langle \boldsymbol{A}_0''(\boldsymbol{S})\Psi_1, \boldsymbol{V}_1\rangle, \Phi_1^*)_{L^2}.$$

This requires the calculation of $\boldsymbol{V}_1$. The function $\boldsymbol{V}_1$ is defined by

$$-\boldsymbol{L}_0\boldsymbol{V}_1 = \frac{1}{2}\boldsymbol{A}_0''(\boldsymbol{S})\Psi_1^2 - \partial_{p_1}\{\tau(\boldsymbol{p})\Psi_1\}|_{\boldsymbol{p}=\boldsymbol{0}}$$

for $\boldsymbol{x} = (x, y)$. Since

$$-\partial_{p_1}\{\tau(\boldsymbol{p})\Psi_1\}|_{\boldsymbol{p}=\boldsymbol{0}} = \begin{pmatrix} \partial_x\{\phi_0(|\boldsymbol{x}|)\cos\theta\} \\ \boldsymbol{0} \end{pmatrix} = \begin{pmatrix} \cos^2\theta\phi_0'(r) + \frac{\sin^2\theta}{r}\phi_0(r) \\ \boldsymbol{0} \end{pmatrix}$$

and the equation of $\boldsymbol{V}_1$ for $\boldsymbol{V}_1 = (v_1, \boldsymbol{q}_1)$ is

$$(-\boldsymbol{L}_0\boldsymbol{V}_1) = -\begin{pmatrix} (D\Delta - \alpha + F'(S(r))H(r_0 - r)v_1 + F(S_0)\delta(r_0 - r)\langle \boldsymbol{e}(\theta), \boldsymbol{q}_1\rangle \\ \beta^*\Gamma'(S_0)\left[\pi r_0 S'\boldsymbol{q}_1 + \int_C v_1\boldsymbol{n}ds\right] \end{pmatrix}$$
$$= \begin{pmatrix} (\cos^2\theta\phi_0' + \frac{\sin^2\theta}{r}\phi_0) + \frac{1}{2}F''(S(r))H(r_0 - r)\cos^2\theta\phi_0^2(r) \\ \frac{1}{2}\beta^*F''(S_0)\int_C \cos^2\theta\phi_0^2(r)\boldsymbol{n}ds \end{pmatrix}$$

$$= \begin{pmatrix} (\cos^2\theta\phi_0' + \frac{\sin^2\theta}{r}\phi_0) + \frac{1}{2}F''(S(r))H(r_0 - r)\cos^2\theta\phi_0^2(r) \\ 0 \end{pmatrix}$$

$$= \frac{1}{2}\begin{pmatrix} \phi_0' + \frac{1}{r}\phi_0 + \frac{1}{2}F''(S(r))H(r_0 - r)\phi_0^2(r) \\ 0 \end{pmatrix}$$

$$+ \frac{1}{2}\cos 2\theta \begin{pmatrix} \phi_0' - \frac{1}{r}\phi_0 + \frac{1}{2}F''(S(r))H(r_0 - r)\phi_0^2(r) \\ 0 \end{pmatrix},$$

$\boldsymbol{V}_1$ is uniquely solved as $\boldsymbol{V}_1(r, \theta) = \begin{pmatrix} Z_0(r) + \cos 2\theta Z_2(r) \\ 0 \end{pmatrix}$ with $\boldsymbol{V}_1 \in E^\perp$. Substituting this $\boldsymbol{V}_1$, we have

$$(\langle \boldsymbol{A}_0''(\boldsymbol{S})\Psi_1, \boldsymbol{V}_1\rangle, \Phi_1^*)_{L^2} = \pi M_1'', \tag{3.30}$$

where

$$M_1'' = \int_0^{r_0} rF''(S(r))\phi_0(r)\left(Z_0(r) + \frac{1}{2}Z_2(r)\right)S'(r)dr$$
$$- \frac{\Gamma''(S(r_0))F(S(r_0))\phi_0(r_0)}{4\Gamma'(S(r_0))}\left(Z_0(r_0) + \frac{1}{2}Z_2(r_0)\right).$$

Finally, we calculate $(\partial_{p_1}\{\tau(\boldsymbol{p})\boldsymbol{V}_1\}|_{\boldsymbol{p}=\boldsymbol{0}}, \Phi_1^*)_{L^2}$. Since $\partial_{p_1}\{\tau(\boldsymbol{p})\boldsymbol{V}_1\}|_{\boldsymbol{p}=\boldsymbol{0}} = -\partial_x \boldsymbol{V}_1$ and

$$\partial_x \boldsymbol{V}_1 = \cos\theta\partial_r \boldsymbol{V}_1 + \frac{\sin\theta}{r}\partial_\theta \boldsymbol{V}_1 = \begin{pmatrix} \cos\theta Z_0' + \cos\theta\cos 2\theta Z_2' - \frac{2\sin\theta\sin 2\theta}{r}Z_2 \\ 0 \end{pmatrix}$$

we can calculate $(\partial_{p_1}\{\tau(\boldsymbol{p})\boldsymbol{V}_1\}|_{\boldsymbol{p}=\boldsymbol{0}}, \Phi_1^*)_{L^2}$ as

$$\begin{aligned} (\partial_{p_1}\{\tau(\boldsymbol{p})\boldsymbol{V}_1\}|_{\boldsymbol{p}=\boldsymbol{0}}, \Phi_1^*)_{L^2} &= -(\partial_x \boldsymbol{V}_1, \Phi_1^*)_{L^2} \\ &= -\frac{1}{2}\pi\int_0^{r_0}(2rZ_0'(r) + rZ_2'(r) + 2Z_2(r))S'(r)dr \\ &=: \pi M_1''' \end{aligned} \tag{3.31}$$

Thus, the coefficient $(\Psi_1, \Phi_1^*)_{L^2}M_1$ is given, making use of (3.29)–(3.31), by

$$(\Psi_1, \Phi_1^*)_{L^2}M_1 = -\pi\left(\frac{1}{8}M_1' + M_1'' - M_1'''\right).$$

Since $(\Psi_1, \Phi_1^*)_{L^2} = \frac{1}{2}(S', \phi_0)_R$ from Lemma 3.7, M_1 is given by

$$M_1 = -\frac{2\pi}{(S', \phi_0)_R}\left(\frac{1}{8}M_1' + M_1'' - M_1'''\right),$$

which proves the lemma for the constant M_1.

M_2 is obtained similarly. Since $\boldsymbol{G}(\boldsymbol{S}) = 0$, we see that $\boldsymbol{V}_2 = 0$ and hence

$$M_2 = -\frac{2}{(S', \phi_0)_R}(\boldsymbol{G}'(\boldsymbol{S})\Psi_1, \Phi_1^*)_{L^2}.$$

$M_2' = (\boldsymbol{G}'(\boldsymbol{S})\Psi_1, \Phi_1^*)_{L^2}$ is

$$\begin{aligned} M_2' &= \left(\begin{pmatrix} 0 & 0 \\ \Gamma'(S(r_0)) \int_C \cdot \boldsymbol{n} ds & \Gamma'(S(r_0))\pi r_0 S'(r_0) \end{pmatrix}\begin{pmatrix} \phi_0(r)\cos\theta \\ \boldsymbol{0} \end{pmatrix}, \Phi_1^*\right)_{L^2} \\ &= \Gamma'(S(r_0))\left\langle \int_C \phi_0(r)\cos\theta \boldsymbol{n} ds, \left(-\frac{F(S(r_0))}{\beta^*\Gamma'(S(r_0))}\right)\boldsymbol{e}_1 \right\rangle \\ &= -\frac{\phi_0(r_0)F(S(r_0))}{\beta^*}, \end{aligned}$$

which proves this lemma for M_2. □

Based on the theorems above, let us discuss the solution of the MB model. Theorem 3.8 implies that the solution $\boldsymbol{U}(t, \boldsymbol{x})$ stays close to $\boldsymbol{U}^*(\boldsymbol{p}(t), \boldsymbol{v}(t))$. The function $\boldsymbol{U}^*(\boldsymbol{p}(t), \boldsymbol{v}(t))$ consists of two terms: a translation of $S(\boldsymbol{x})$ and a perturbation in the direction of two generalized eigenfunctions. The function $\boldsymbol{v}(t)$ represents the deformation of the solution from the radially symmetric stationary solution $S(\boldsymbol{x})$. In addition, it has another meaning that it corresponds to the velocity of the camphor disk because it approximates $\dot{\boldsymbol{p}}(t)$ as can be seen from (3.26).

The analysis of camphor motion is reduced to (3.26). The property of (3.26) depends on the signs of the coefficients M_1 and M_2. First, the stability of the radially symmetric stationary solution $S(\boldsymbol{x}, \boldsymbol{p})$ depends on the sign of M_2. If M_2 is negative, then $\boldsymbol{S}(\boldsymbol{x}, \boldsymbol{p})$ is stable for $\eta < 0$ ($\beta < \beta^*$) and unstable for $\eta > 0$ ($\beta > \beta^*$). Conversely, if M_2 is positive, then $\boldsymbol{S}(\boldsymbol{x}, \boldsymbol{p})$ is unstable for $\eta < 0$ and stable for $\eta > 0$. Since we have $\phi_0(r) \leq 0$ and $F(c) \geq 0$, we can conclude that M_2 is negative. Second, the direction of bifurcation depends on the sign of M_1. The bifurcation is supercritical if $M_1 > 0$ and subcritical if $M_1 < 0$. If $M_1 > 0$ and $M_2 < 0$, then the bifurcation diagram is as shown in Fig. 1.16. Unfortunately, the sign of M_1 has not yet been theoretically determined. Numerical simulations such as Fig. 1.7 strongly suggest that M_1 is positive. In the following chapters, we continue the discussion by assuming $M_1 > 0$ and $M_2 < 0$.

The ODE (3.26) describing the motion of a disk is reduced to (1.8) by truncating higher order terms. Thus, we have obtained the particle model on the entire plane from the MB model.

3.4 Repulsive Interaction of Two Disks

In this section, we derive the equation describing the interaction of two disks. Portions of this section are adapted from [9], used with permission of the American Institute of Mathematical Sciences, from Chen et al. [9]; permission conveyed through Copyright Clearance Center, Inc.

In the Introduction, we observed that a moving disk reflects from a wall, as in Fig. 1.2. Our strategy to understand this property is based upon the fact that two moving spots which are positioned in a mirror symmetric fashion with respect to a straight wall repel each other when they are near as in Fig. 3.5. For this reason, we consider the interaction of two moving spots in $\mathbb{R}^2$. The moving boundary model for multiple disks is

$$\begin{cases} c_t = D\Delta c - \alpha c + F(c)\sum_{j=1}^{m} H(r_j - |\boldsymbol{x} - \boldsymbol{p}_j|), & x \in \mathbb{R}^2, t \in (0,\infty), \\ \dot{\boldsymbol{p}}_j(t) = \beta_j \int_{C_{r_j}} \Gamma(c(\boldsymbol{p}_j + \boldsymbol{y}))\boldsymbol{n} ds, & t > 0, j = 1, \dots, m, \end{cases} \tag{3.32}$$

where C_{r_j} is defined by $C_{r_j} = B(\boldsymbol{0}, r_j)$. In accordance with what we have just explained regarding reflections from a wall, we now only consider the case of two disks ($m = 2$). Let $\boldsymbol{S}_1(\boldsymbol{x}; \boldsymbol{p}_1), \boldsymbol{S}_2(\boldsymbol{x}; \boldsymbol{p}_2) \in \mathbb{R}^1 \times \mathbb{R}^2 \times \mathbb{R}^2$ be given by

$$\boldsymbol{S}_1(\boldsymbol{x}; \boldsymbol{p}_1) = \begin{pmatrix} S_1(\boldsymbol{x}) \\ \boldsymbol{p}_1 \\ \boldsymbol{0} \end{pmatrix}, \quad \boldsymbol{S}_2(\boldsymbol{x}; \boldsymbol{p}_2) = \begin{pmatrix} S_2(\boldsymbol{x}) \\ \boldsymbol{0} \\ \boldsymbol{p}_2 \end{pmatrix},$$

where $S_j(\boldsymbol{x})$ are the radially symmetric solutions with radii $r = r_j$ $(j = 1, 2)$ discussed in Sect. 2.3. We assume the asymptotic forms

$$S_j(r) \to \frac{1}{\sqrt{r}} e^{-\alpha r} b_j \ (r \to \infty).$$

We write (3.32) ($m = 2$) in terms of $\boldsymbol{y} = \boldsymbol{x} - \boldsymbol{p}_1$, $\boldsymbol{d} = \boldsymbol{p}_2 - \boldsymbol{p}_1$ and $c = c(t, \boldsymbol{y})$

$$\begin{cases} c_t = D\Delta c - \alpha c + F(c)\{H(r_1 - r) + H(r_2 - |\boldsymbol{y} - \boldsymbol{d}|)\} + \langle \dot{\boldsymbol{p}}_1, \nabla c\rangle, \\ \dot{\boldsymbol{p}}_1(t) = \beta_1 \displaystyle\int_{C_{r_1}} \Gamma(c(\boldsymbol{y}))\boldsymbol{n} ds, \\ \dot{\boldsymbol{p}}_2(t) = \beta_2 \displaystyle\int_{C_{r_2}} \Gamma(c(\boldsymbol{y}+\boldsymbol{d}))\boldsymbol{n} ds, \end{cases} \tag{3.33}$$

or simply

$$\boldsymbol{U}_t = \boldsymbol{A}(\boldsymbol{U}) + \begin{pmatrix} \langle \dot{\boldsymbol{p}}_1, \nabla c\rangle \\ \boldsymbol{0} \\ \boldsymbol{0} \end{pmatrix}$$

for $\boldsymbol{U} = (c, \boldsymbol{p}_1, \boldsymbol{p}_2) \in \mathbb{R}^1 \times \mathbb{R}^2 \times \mathbb{R}^2$. We here assume β_1 is close to the bifurcation point, say β_1^* as in the previous section, while we make no such assumption for β_2. Let $\beta_1 = \beta_1^* + \eta$ and write $\boldsymbol{A} = \boldsymbol{A}_0 + \eta \boldsymbol{G}$.

We only consider the dynamics of $\boldsymbol{p}_1$ and $\boldsymbol{v}_1$ and hereafter in this section, and we simply express S_1, β_1, r_1 and C_{r_1} in terms of S, β, r_0, C and so on. Then, the dynamics of $\boldsymbol{p} = \boldsymbol{p}_1 = (p_1, p_2)$ and $\boldsymbol{v} = \boldsymbol{v}_1$ are basically given by

$$\begin{cases} \dot{\boldsymbol{p}} = \boldsymbol{v} + \dfrac{2}{(S', \phi_0)_{L^2}} \begin{pmatrix} (\boldsymbol{A}_0(\boldsymbol{S}(\boldsymbol{y}; \boldsymbol{p}) + \boldsymbol{S}_2(\boldsymbol{y} - \boldsymbol{d}; \boldsymbol{p}_2)), \Psi_1^*)_{L^2} \\ (\boldsymbol{A}_0(\boldsymbol{S}(\boldsymbol{y}; \boldsymbol{p}) + \boldsymbol{S}_2(\boldsymbol{y} - \boldsymbol{d}; \boldsymbol{p}_2)), \Psi_2^*)_{L^2} \end{pmatrix} + h.o.t., \\ \dot{\boldsymbol{v}} = -\nabla_{\boldsymbol{v}} W + \dfrac{2}{(S', \phi_0)_{L^2}} \begin{pmatrix} (\boldsymbol{A}_0(\boldsymbol{S}(\boldsymbol{y}; \boldsymbol{p}) + \boldsymbol{S}_2(\boldsymbol{y} - \boldsymbol{d}; \boldsymbol{p}_2)), \Phi_1^*)_{L^2} \\ (\boldsymbol{A}_0(\boldsymbol{S}(\boldsymbol{y}; \boldsymbol{p}) + \boldsymbol{S}_2(\boldsymbol{y} - \boldsymbol{d}; \boldsymbol{p}_2)), \Phi_2^*)_{L^2} \end{pmatrix} + h.o.t., \end{cases} \tag{3.34}$$

which is derived by applying Theorems 3 and 4 in [16]. Here, Ψ_1^* and Φ_1^* are given by

$$\Psi_1^* = \begin{pmatrix} (-a_0 S'(r) + \phi_0(r)) \cos\theta \\ \frac{a_0 F(S_0)}{\beta^* \Gamma'(S_0)} \boldsymbol{e}_1 \\ \boldsymbol{0} \end{pmatrix} \in \mathbb{R}^1 \times \mathbb{R}^2 \times \mathbb{R}^2,$$

and

$$\Phi_1^* = \begin{pmatrix} S'(r) \cos\theta \\ -\frac{F(S_0)}{\beta^* \Gamma'(S_0)} \boldsymbol{e}_1 \\ \boldsymbol{0} \end{pmatrix} \in \mathbb{R}^1 \times \mathbb{R}^2 \times \mathbb{R}^2$$

using the same symbols as in Lemma 3.7 without ambiguity. $W = W(\boldsymbol{v})$ is the function in Theorem 3.8.

In this section, we assume (3.34) and derive the explicit form of (3.34) for Eq. (3.33).

Lemma 3.10 *Let $d = |\boldsymbol{d}|$ and $\boldsymbol{e} = (e_1, e_2) = \frac{1}{d}(\boldsymbol{p}_2 - \boldsymbol{p})$. Then*

$$(\boldsymbol{A}_0(\boldsymbol{S}(\boldsymbol{y}; \boldsymbol{p}) + \boldsymbol{S}_2(\boldsymbol{y} - \boldsymbol{d}; \boldsymbol{p}_2)), \Psi_j^*)_{L^2} = \frac{b_2}{\sqrt{d}} e^{-\alpha d} M_0 e_j (1 + o(1)) \ (j = 1, 2),$$

$$(\boldsymbol{A}_0(\boldsymbol{S}(\boldsymbol{y};\boldsymbol{p}) + \boldsymbol{S}_2(\boldsymbol{y}-\boldsymbol{d};\boldsymbol{p}_2)), \Phi_j^*)_{L^2} = \frac{b_2}{\sqrt{d}} e^{-\alpha d} \overline{M_0} e_j (1+o(1)) \ (j=1,2),$$

hold, where

$$M_0 = a_0 F(S_0) r_0 \int_0^{2\pi} e^{\alpha r_0 \cos\theta} \cos\theta d\theta + \int_0^{r_0} G_0(r) \int_0^{2\pi} e^{\alpha r \cos\theta} \cos\theta d\theta dr$$

with $G_0(r) = \{F'(S(r)) - F'(0)\}(-a_0 S'(r) + \phi_0(r))$, *and*

$$\overline{M_0} = a_0 F(S_0) r_0 \int_0^{2\pi} e^{\alpha r_0 \cos\theta} \cos\theta d\theta + \int_0^{r_0} G_1(r) \int_0^{2\pi} e^{\alpha r \cos\theta} \cos\theta d\theta dr$$

with $G_1(r) = \{F'(S(r)) - F'(0)\} S'(r)$.

Proof Since the proof of this lemma is quite similar to the proof of Theorem 4.2 [16], we just give the outline by proving the first equation of this lemma

$$(\boldsymbol{A}_0(\boldsymbol{S}(\boldsymbol{y};\boldsymbol{p}) + \boldsymbol{S}_2(\boldsymbol{y}-\boldsymbol{d};\boldsymbol{p}_2)), \Psi_j^*)_{L^2} = \frac{b_2}{\sqrt{d}} e^{-\alpha d} M_0 e_1 (1+o(1)),$$

while we omit the others.

Let $\boldsymbol{A}_0 = (\mathscr{L}_1(c), \mathscr{L}_2(c), \mathscr{L}_3(c))$. Then the left hand side of the equation above is

$$\begin{aligned}&(\mathscr{L}_1(S(\boldsymbol{y}) + S_2(\boldsymbol{y}-\boldsymbol{d})), (-a_0 S' + \phi_0)\cos\theta)_{L^2} \\ &\quad + \frac{a_0 F(S(r_0))}{\Gamma'(S(r_0))} \int_C \Gamma(S(\boldsymbol{y}) + S_2(\boldsymbol{y}-\boldsymbol{d})) \cos\theta ds,\end{aligned}$$

where $\boldsymbol{y} = r_0(\cos\theta, \sin\theta)$.

First, we calculate

$$\frac{a_0 F(S_0)}{\Gamma'(S_0)} \int_C \Gamma(S(\boldsymbol{y}) + S_2(\boldsymbol{y}-\boldsymbol{d})) \cos\theta ds. \tag{3.35}$$

Since

$$S_2(\boldsymbol{y}-\boldsymbol{d}) \sim \frac{1}{\sqrt{|\boldsymbol{y}-\boldsymbol{d}|}} e^{-\alpha|\boldsymbol{y}-\boldsymbol{d}|} b_2 \sim \frac{1}{\sqrt{d}} e^{-\alpha d} e^{-\alpha r \cos(\theta-\gamma)} b_2$$

for $e_1 = \cos\gamma$, we substitute this into (3.35) and we have

$$
\begin{aligned}
&\frac{a_0 F(S_0)}{\Gamma'(S_0)} \left(\int_{C_{r_0}} \Gamma(S(\mathbf{y})) \cos\theta ds + \int_C \Gamma'(S(\mathbf{y})) S_2(\mathbf{y}-\mathbf{d}) \cos\theta ds \right) \\
&\sim a_0 F(S_0) \int_C S_2(\mathbf{y}-\mathbf{d}) \cos\theta ds \\
&\sim a_0 F(S_0) \frac{1}{\sqrt{d}} e^{-\alpha d} b_2 r_0 e_1 \int_0^{2\pi} e^{\alpha r_0 \cos\theta} \cos\theta d\theta.
\end{aligned}
$$

Here $f \sim g$ denotes $f = (1 + o(1))g$ as $d \to \infty$. Thus,

$$
\begin{aligned}
&\frac{a_0 F(S_0)}{\Gamma'(S_0)} \int_C \Gamma(S(\mathbf{y}) + S_2(\mathbf{y}-\mathbf{d})) \cos\theta ds \\
&\sim \frac{b_2}{\sqrt{d}} e^{-\alpha d} a_0 F(S(r_0)) r_0 e_1 \int_0^{2\pi} e^{\alpha r_0 \cos\theta} \cos\theta d\theta
\end{aligned}
\tag{3.36}
$$

holds.

Second, computing

$$
(\mathcal{L}_1(S(y) + S_2(\mathbf{y}-\mathbf{d})), (-a_0 S' + \phi_0)\cos\theta)_{L^2},
$$

we find

$$
\begin{aligned}
&(\mathcal{L}_1(S(\mathbf{y}) + S_2(\mathbf{y}-\mathbf{d})), (-a_0 S' + \phi_0)\cos\theta)_{L^2} \\
&\sim \Big(\{F'(S(r)) - F'(0)\}\{H(r_0 - r) + H(r_2 - |\mathbf{y}-\mathbf{d}|)\} \\
&\qquad \frac{1}{\sqrt{d}} e^{-\alpha d} e^{\alpha r \cos(\theta-\gamma)} b_2, (-a_0 S' + \phi_0)\cos\theta \Big)_{L^2} \\
&\sim \frac{b_2}{\sqrt{d}} e^{-\alpha d} \int_0^{r_0} r\{F'(S(r)) - F'(0)\}(-a_0 S'(r) + \phi_0(r)) \\
&\qquad \int_0^{2\pi} e^{\alpha r \cos\theta} \cos(\theta + \gamma) d\theta dr \\
&= \frac{b_2}{\sqrt{d}} e^{-\alpha d} e_1 \int_0^{r_0} G_0(r) \int_0^{2\pi} e^{\alpha r \cos\theta} \cos\theta d\theta dr,
\end{aligned}
$$

where $G_0(r) = r\{F'(S(r)) - F'(0)\}(-a_0 S'(r) + \phi_0(r))$. This and (3.36) suffice to prove this lemma. □

Remark 3.11 For a simple example of F such as $F(c) = F_0 - d_0 c$ for positive constants F_0 and d_0, $G_0(r) = G_1(r) = 0$ hold and the coefficients M_0, $\overline{M_0}$ can be simply shown to be positive.

If two camphor disks are identical, that is, $r_1 = r_2 =: r_0$ and $\beta_2 = \beta_1 =: \beta^* + \eta$ in (3.32), then by (3.34) and Lemma 3.10, the movement of two interacting disks is essentially described as follows:

$$\begin{cases} \dot{\boldsymbol{p}}_1 = \boldsymbol{v}_1 - \dfrac{M_0^*}{\sqrt{d}} e^{-\alpha d} \boldsymbol{e}, \\ \dot{\boldsymbol{v}}_1 = -\nabla_v W - \dfrac{\overline{M_0^*}}{\sqrt{d}} e^{-\alpha d} \boldsymbol{e}, \\ \dot{\boldsymbol{p}}_2 = \boldsymbol{v}_2 + \dfrac{M_0^*}{\sqrt{d}} e^{-\alpha d} \boldsymbol{e}, \\ \dot{\boldsymbol{v}}_2 = -\nabla_v W + \dfrac{\overline{M_0^*}}{\sqrt{d}} e^{-\alpha d} \boldsymbol{e}, \end{cases} \tag{3.37}$$

where

$$\boldsymbol{e} = (e_1, e_2) = \frac{1}{d}(\boldsymbol{p}_2 - \boldsymbol{p}_1),$$

$$M_0^* = -\frac{2}{(S', \phi_0)_{L^2}} b_0 M_0, \quad \overline{M_0^*} = -\frac{2}{(S', \phi_0)_{L^2}} b_0 \overline{M_0},$$

and $b_0 = b_1 = b_2$. Here we note that (3.37) includes the lowest order terms only and the higher order terms have been truncated.

Remark 3.12 Equation 3.37 is universal in some sense because similar types of equations to (3.37) appear commonly in different fields (see e.g. [45]). It strongly suggests that the essential structures for the dynamics are captured by (3.37), at least in the neighborhood of the bifurcation point. The specific details of model equations merely appear in the specific values of coefficients. Other differences between model equations may appear in higher order terms. In practice, (3.37) is the lowest order equation and the truncated higher order terms are $O(|\boldsymbol{v}_1|^3 + |\boldsymbol{v}_2|^3 + |\eta|^{\frac{3}{2}})$ for $\dot{\boldsymbol{p}}_j$ and $O(|\boldsymbol{v}_1|^4 + |\boldsymbol{v}_2|^4 + |\eta|^2)$ for $\dot{\boldsymbol{v}}_j$ (Theorem 3 in [16]). The explicit forms of these higher order terms are prohibitively difficult to compute and we make no further comment about them here.

If two disks are in mirror-symmetric positions with respect to the y-axis in $\mathbb{R}^2$, then for $\boldsymbol{p}_1 = (x, y)$ and $\boldsymbol{v}_1 = (v, w)$, $\boldsymbol{p}_2$ and $\boldsymbol{v}_2$ are respectively given by $\boldsymbol{p}_2 = (-x, y)$ and $\boldsymbol{v}_2 = (-v, w)$, which allows for (3.37) to be written in the simpler form

$$\begin{cases} \dot{x} = v + \dfrac{M_0^*}{\sqrt{2x}} e^{-2\alpha x}, \\ \dot{y} = w, \\ \dot{\boldsymbol{v}}_1 = -\nabla_v W + \dfrac{\overline{M_0^*}}{\sqrt{2x}} e^{-2\alpha x} \begin{pmatrix} 1 \\ 0 \end{pmatrix}. \end{cases} \tag{3.38}$$

Figures 1.13 and 1.17 are numerical simulations of (3.32) (for $m = 2$) and (3.38). Thus, the reflection behavior observed in real and numerical experiments (Figs. 1.2, 3.4 and 3.5) is theoretically reproduced by the reduced Eq. 3.38. A detailed analysis of (3.38) is presented in the next chapter.

3.5 Some Remarks

The reduction theory presented in this chapter is based on [9, 13, 16] for smooth nonlinearity. However, the MB model involves Dirac δ-function coefficients as in (3.18). We note that all the results in this chapter can be proven by approximating such δ-function coefficients with smooth functions when we directly deal with the MB model owing to the simplicity of explicit calculations. Recently, it was reported [30] that the direct treatment of δ-function coefficients is possible in a mathematically rigorous manner.

The interaction of localized patterns in reaction-diffusion systems has been extensively studied. Kawasaki and Ohta [35] have proposed deriving ordinary differential equations that approximate the dynamics and interaction of traveling fronts in one-dimensional reaction-diffusion systems from the viewpoint of nonlinear physics. Such a methodology for deriving equations of motion has been extended and applied to localized patterns such as fronts and pulses in 1D and spots in 2D etc. [16, 17]. See [9, 16], and references therein, for a mathematical justification and a systematic derivation.

Although the reduction theory framework is valid for more general equations, it is difficult to determine the coefficients in the reduced equations because of the difficulty of adjoint problems. Typical examples whose adjoint eigenvalue problems are explicitly solved are the FitzHugh–Nagumo equation, the Gray–Scott model, and the Gierer–Meinhardt model. Indeed, the repulsive interaction between two localized pulses and spots has been investigated in detail (e.g., [13, 18]). The moving boundary model is also capable of explicit calculation, to some extent, as shown in this chapter (also refer to [14] for a nonradial MB model). However, it is difficult to determine the sign of M_1. As the critical eigenvalue is algebraically degenerate with a Jordan block, it is also a challenging problem to numerically compute the associated eigenfunctions and hence the coefficients.

In this chapter and throughout the book, we only consider the case where the repulsive interaction is weak, i.e., the case where two camphor disks are sufficiently far apart from each other. While the theoretical framework is applicable to other dissipative systems as mentioned above, there are many other possible behaviors of interacting spots beyond weak interaction, including collision, annihilation, merging, trapping, and interactions with obstacles. These are the cases of strong interaction (refer to e.g., [38, 45, 62–65, 81]).

In the following sections, we mainly deal with the weak interaction of camphor disks, whereas reflection from a boundary is treated using the reduced ODE system (3.38). The difference between the angles of incidence and reflection plays a particularly important role. Similar phenomena appear in the refraction and reflection of particles in various other systems, including dissipative systems and optics, as described in [64]. Snell's law is an example, which describes refraction at interfaces in inhomogeneous media. Here, we have not provided any explicit law for the relationship between the angles of incidence and reflection, which requires careful analysis of the ODE system (3.38).

Chapter 4
Particle Models

Abstract Particle models are considered for the entire plane, half-plane, and rectangular domains. It is proved that the angle of reflection is greater than that of incidence under certain assumptions. The bifurcation of periodic orbits in a rectangular domain is studied analytically and numerically.

4.1 Reflection from the Boundary

To better understand the reflections of a camphor disk, we now return to exploring the particle model. Portions of this section are adapted from [15], used with permission of the American Institute of Mathematical Sciences, from Ei et al. [15]; permission conveyed through Copyright Clearance Center, Inc.

Recall that the particle model (3.37) describes two disks repulsively interacting with each other:

$$\begin{cases} \dot{\boldsymbol{p}}_1 = \boldsymbol{v}_1 - m_0 \tilde{h}(d) \dfrac{\boldsymbol{p}_2 - \boldsymbol{p}_1}{d}, \\ \dot{\boldsymbol{v}}_1 = \left(\delta - |\boldsymbol{v}_1|^2\right) \boldsymbol{v}_1 - m_2 \tilde{h}(d) \dfrac{\boldsymbol{p}_2 - \boldsymbol{p}_1}{d}, \\ \dot{\boldsymbol{p}}_2 = \boldsymbol{v}_2 - m_0 \tilde{h}(d) \dfrac{\boldsymbol{p}_1 - \boldsymbol{p}_2}{d}, \\ \dot{\boldsymbol{v}}_2 = \left(\delta - |\boldsymbol{v}_2|^2\right) \boldsymbol{v}_2 - m_2 \tilde{h}(d) \dfrac{\boldsymbol{p}_1 - \boldsymbol{p}_2}{d}, \end{cases} \tag{4.1}$$

where $d = |\boldsymbol{p}_2 - \boldsymbol{p}_1|$, $\tilde{h}(x) = e^{-x}/\sqrt{x}$, $m_0 \geq 0$, $m_2 > 0$ and $\delta > 0$. It is easy to see that this system is symmetrical [9, 16]:

Lemma 4.1 *Consider a solution* $(\boldsymbol{p}_1, \boldsymbol{v}_1, \boldsymbol{p}_2, \boldsymbol{v}_2)$ *of* (4.1). *Let* $\boldsymbol{p}_1(t) = (x(t), y(t))$ *and* $\boldsymbol{v}_1(t) = (v(t), w(t))$. *If* $\boldsymbol{p}_2(0) = (-x(0), y(0))$ *and* $\boldsymbol{v}_2(0) = (-v(0), w(0))$, *then* $\boldsymbol{p}_2(t) = (-x(t), y(t))$ *and* $\boldsymbol{v}_2(t) = (-v(t), w(t))$ *hold for any* $t > 0$.

T. Miyaji et al., *A Billiard Problem in Nonlinear Dissipative Systems*,
Surveys and Tutorials in the Applied Mathematical Sciences 18,
https://doi.org/10.1007/978-981-95-7561-9_4

Proof We have $\dot{\boldsymbol{p}}_1 + \dot{\boldsymbol{p}}_2 = \boldsymbol{v}_1 + \boldsymbol{v}_2$ and

$$\dot{\boldsymbol{v}}_1 + \dot{\boldsymbol{v}}_2 = \left(\delta - |\boldsymbol{v}_1|^2\right)\boldsymbol{v}_1 + \left(\delta - |\boldsymbol{v}_2|^2\right)\boldsymbol{v}_2.$$

If $\boldsymbol{p}_2(0) = (-x(0), y(0))$ and $\boldsymbol{v}_2(0) = (-v(0), w(0))$, then

$$\dot{\boldsymbol{p}}_1(0) + \dot{\boldsymbol{p}}_2(0) = (0, 2w(0))$$

and

$$\dot{\boldsymbol{v}}_1(0) + \dot{\boldsymbol{v}}_2(0) = \left(\delta - v^2(0) - w^2(0)\right)(0, 2w(0)).$$

Therefore $\boldsymbol{p}_2(t) = (-x(t), y(t))$ and $\boldsymbol{v}_2(t) = (-v(t), w(t))$ hold for any $t > 0$. □

Suppose that a particle is on the right half-plane $\{x > 0\}$ and that its mirror image with respect to the y-axis lies on the left half-plane. From the Lemma above, the motion of the particle is governed by the following equations for (x, y, v, w) [9, 16]:

$$\begin{cases} \dot{x} = v + mh(x), \\ \dot{y} = w, \\ \dot{v} = \left(\delta - v^2 - w^2\right)v + h(x), \\ \dot{w} = \left(\delta - v^2 - w^2\right)w, \end{cases} \tag{4.2}$$

where $h(x) = m_2\tilde{h}(2x)$ and $m = m_0/m_2$. The system (4.2) describes the motion of a camphor disk on the right half-plane that interacts with the boundary $\{x = 0\}$ in a repulsive manner. Indeed, $h(x)$ is positive and monotonically decreasing for $x > 0$, and becomes large as x is close to the boundary. Note that the equations for (x, v, w) are independent of y: $y(t)$ is obtained by integrating $\dot{y} = w$ with respect to t. If $\dot{y}(0) = w(0)$ is positive, then $\dot{y}(t) = w(t) > 0$ holds for any $t > 0$. We only consider $w \geq 0$, as the case where $w < 0$ is essentially same as $w > 0$ by symmetry. Indeed, (4.2) is equivariant under $w \mapsto -w$.

In this section, the following two propositions are proved, referring to [15]:

(i) The particle governed by (4.2) reflects from the boundary at most once.
(ii) The angle of reflection is greater than that of incidence.

In advance, let us impose some assumptions. As the particle model has been derived under the assumption that δ is sufficiently close to zero, we assume

$$0 < \delta < \frac{1}{m}.$$

If δ is greater than $1/m$, then (4.2) has nontrivial equilibria. Such cases are excluded from consideration. In [15], $0 < \delta < \infty$ was assumed because the case where $m = 0$ was treated. Here, we make a slightly different assumption on δ as above because we prove (i) for nonnegative m.

In addition, we will deal with more general $h(x)$ than $\exp(-2x)/\sqrt{2x}$. Assume that the function $h(x)$ satisfies the following [15]:

Assumption 4.1 *([15]) The function $h(x)$ is a C^2 function on $(0, \infty)$ which satisfies the following conditions:*

(i) $h(x) > 0$ *for any* $x > 0$,
(ii) there exists $a > 0$ *such that* $h'(x) \leq -ah(x)$ *for any* $x > 0$,
(iii) $h''(x) > 0$ *for any* $x > 0$,
(iv) $h'(x) \to 0$ *as* $x \to \infty$, *and*
(v) there exists $c > 0$ *such that* $h''(x)/h'(x) \to -c$ *as* $x \to +\infty$.

Note that the function $h(x) = m_2 \exp(-2x)/\sqrt{2x}$ satisfies these conditions with $a = 2$ and $c = 2$.

The proof of (i) is based on the phase-space analysis. Since the orbit for (4.2) we will consider is unbounded in the original coordinate system, it is not easy to handle from the viewpoint of dynamical systems. In [15], a new coordinate system is introduced by

$$z = h(x), \quad v = r\cos\theta, \quad w = r\sin\theta. \tag{4.3}$$

In these coordinates, we obtain

$$\begin{cases} \dot{z} = H(z)(r\cos\theta + mz), \\ \dot{r} = (\delta - r^2)\, r + z\cos\theta, \\ \dot{\theta} = -\dfrac{z}{r}\sin\theta, \end{cases} \tag{4.4}$$

where $H(z)$ is defined by

$$H(z) = \begin{cases} \mathrm{sgn}(z)h'(h^{-1}(\mathrm{sgn}(z)z)) & (z \neq 0), \\ 0 & (z = 0). \end{cases}$$

$H(z)$ is well-defined since $h(x)$ satisfies Assumption 4.1.

Definition 4.2 ([15]) Suppose that there exists a solution of (4.4) satisfying

$$\lim_{t \to \pm\infty} (z(t), r(t), \theta(t)) = (0, \sqrt{\delta}, \theta_{\pm\infty}). \tag{4.5}$$

for some $\theta_{-\infty}$ and θ_{∞}. Then the angle of incidence and that of reflection are defined by

$$\theta_{\mathrm{inc}} = \pi - \theta_{-\infty}, \quad \text{and} \quad \theta_{\mathrm{ref}} = \theta_{\infty},$$

respectively.

Figure 4.1 illustrates Definition 4.2. As Theorem 4.3 below proves, for $\theta_{-\infty} \in (\pi/2, \pi)$, there exists a solution of (4.4) with (4.5) for some θ_{∞}. For such a solution,

Fig. 4.1 Definition of the angles θ_{inc} and θ_{ref}. Figure adapted from [15, Fig. 2(B)], used with permission of the American Institute of Mathematical Sciences, from Ei et al. [15]; permission conveyed through Copyright Clearance Center, Inc.

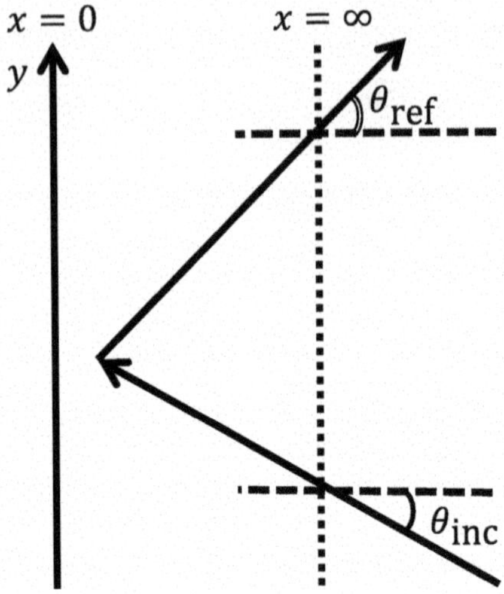

θ_{inc} is the angle formed by vectors $(v, w) = \sqrt{\delta}(\cos\theta_{-\infty}, \sin\theta_{-\infty})$ and $(-1, 0)$, and θ_{ref} is the angle formed by vectors $(v, w) = \sqrt{\delta}(\cos\theta_{\infty}, \sin\theta_{\infty})$ and $(1, 0)$.

Theorem 4.3 *([15]) Suppose that $\theta_{-\infty}$ is in the interval $[\pi/2, \pi]$. The following properties hold:*

1. *There exists a $\theta_\infty \in [0, \pi/2]$ and a solution of (4.4) satisfying (4.5).*
2. *If $\theta_{-\infty} \neq \pi/2$, then $\dot{x}(t)$ changes its sign from negative to positive just once.*
3. *If $m = 0$ holds, then $\sin\theta_{-\infty} \leq \sin\theta_\infty$ holds, with equality only if $\theta_{-\infty} = \pi$ or $\theta_{-\infty} = \pi/2$.*

Accordingly, θ_{ref} is a function of θ_{inc}. Moreover, if $m = 0$, then $\theta_{\mathrm{inc}} \leq \theta_{\mathrm{ref}}$ holds, where equality holds only if $\theta_{\mathrm{inc}} = 0$ or $\pi/2$.

Theorem 4.3 is proved for $m = 0$ in [15]. We present a proof of the first two statements for nonnegative m. The idea of proof is the same as [15], but we have to modify it to consider the case of $m > 0$. It is crucial to show the existence of a heteroclinic orbit. The proof is divided into three cases. (i) $\theta_{-\infty} = \pi/2$, (ii) $\theta_{-\infty} = \pi$ and (iii) $\theta_{-\infty} \in (\pi/2, \pi)$.

Lemma 4.4 *([15]) $H(z)$ is a monotonically decreasing C^1 function on $\mathbb{R}$, $H(z) \leq -az$ holds for $z \geq 0$, and $H'(0) = -c < 0$, where a and c are the constants appearing in Assumption 4.1.*

Proof By Assumption 4.1(iv), the map $h : (0, \infty) \cup \{\infty\} \to [0, \infty)$ is surjective. Since $h^{-1}(z) > 0$ for any $z \geq 0$, we obtain

$$H(z) = h'(h^{-1}(z)) \leq -ah(h^{-1}(z)) = -az.$$

The derivative $H'(z)$ is given by

$$H'(z) = \frac{h''(h^{-1}(z))}{h'(h^{-1}(z))}$$

for $z > 0$. By Assumption 4.1(iv) and (v), we obtain

$$\lim_{z\to+0} H'(z) = -c.$$

This proves that $H'(z)$ is a C^1 function on $\mathbb{R}$. □

It should be noted that $H(z)$ is not necessarily C^2 at $z = 0$. We then have

$$H''(h(x)) = -\frac{(h''(x))^2}{(h'(x))^3} + \frac{h'''(x)}{(h'(x))^2}.$$

If $h(x) = (2x)^{-\frac{1}{2}}\exp(-2x)$, then

$$H''(h(x)) = -\frac{\sqrt{2}(3 + 24x + 16x^2)}{\sqrt{x}(1 + 4x)^3} e^{2x} \to -\infty$$

when $x \to \infty$.

The simplest case is when no reflection occurs. This is obvious because $(z, r, \theta) = (0, \sqrt{\delta}, \pi/2)$ is an equilibrium of (4.4).

Lemma 4.5 *([15]) Let $\theta_{-\infty} = \pi/2$. Then $(z(t), r(t), \theta(t)) = (0, \sqrt{\delta}, \pi/2)$ is a solution of* (4.4) *satisfying* (4.5) *with $\theta_\infty = \pi/2$, and $\sin\theta_\infty = \sin\theta_{-\infty} = 1$ holds.*

In the other simplest case, (4.2) is reduced to a planar vector field.

Lemma 4.6 *([15]) Let $\theta_{-\infty} = \pi$. Then there exists a heteroclinic solution of* (4.2) *from $(z, v, w) = (0, -\sqrt{\delta}, 0)$ to $(0, \sqrt{\delta}, 0)$. In this case $\theta_\infty = 0$ and $\sin\theta_\infty = \sin\theta_{-\infty} = 0$.*

Proof The proof is the same as [15] and uses the Poincaré–Bendixson theorem, but we make some modifications to consider nonnegative m. In this case, the trajectory is perpendicular to the boundary. Indeed, the set $\{(z, v, w) \mid z \geq 0, w = 0\}$ is invariant for (4.2). It is sufficient to consider the vector field restricted to the (z, v)-plane, which is given by

$$\dot{z} = H(z)(v + mz), \quad \dot{v} = \left(\delta - v^2\right)v + z. \tag{4.6}$$

As our aim is to show that there is a heteroclinic orbit, we have to find equilibria. This system has three equilibria:

$$(z, v) = (0, -\sqrt{\delta}), \quad (0, 0), \quad (0, \sqrt{\delta}).$$

Let us consider the linearized eigenvalues of these equilibria. The Jacobian matrix of (4.6) is given by

$$J(z,v) = \begin{pmatrix} H'(z)(v+mz) + mH(z) & H(z) \\ 1 & \delta - 3v^2 \end{pmatrix}.$$

We obtain

$$J(0, \pm\sqrt{\delta}) = \begin{pmatrix} \mp c\sqrt{\delta} & 0 \\ 1 & -2\delta \end{pmatrix}, \quad J(0,0) = \begin{pmatrix} 0 & 0 \\ 1 & \delta \end{pmatrix},$$

and hence,

$$\sigma(J(0, \pm\sqrt{\delta})) = \left\{-2\delta, \mp c\sqrt{\delta}\right\}, \quad \sigma(J(0,0)) = \{0, \delta\},$$

where $\sigma(A)$ denotes the set of all eigenvalues for a matrix A. Since δ and c are positive, $(0, \sqrt{\delta})$ is a stable node and $(0, -\sqrt{\delta})$ is a saddle. The origin has a one-dimensional unstable manifold and a one-dimensional center manifold. As the center manifold is tangential to span$\{(-\delta, 1)\}$ at the origin, it intersects $\{(z,v) \mid v + mz < 0, z > 0\}$ in which $\dot{z} > 0$ holds. Therefore, the origin is not an ω-limit set of any initial point in $\{(z,v) \mid z > 0\}$.

The idea of proof is as follows [15]. Take an initial value $(z_0, -\sqrt{\delta})$ with $\delta^{\frac{3}{2}} < z_0 < \sqrt{\delta}/m$ and show that $\omega(z_0, -\sqrt{\delta}) = \{(0, \sqrt{\delta})\}$. Let D be the closed region bounded by the following four curves:

$$\{(z,v) \mid 0 \le z \le z_0, v = -\sqrt{\delta}\},$$
$$\{(z,v) \mid z = v(v^2 - \delta), -\sqrt{\delta} \le v \le 0\},$$
$$\{(z,v) \mid z = 0, 0 \le v \le \sqrt{\delta}\},$$

and the closure of the positive orbit of $(z_0, -\sqrt{\delta})$. D is positively invariant and contains no cycles. The Poincaré–Bendixson theorem implies that, for any point in the interior of D, its ω-limit set is $\{(0, \sqrt{\delta})\}$. Note that the origin cannot be an attractor. Therefore, in particular, the one-dimensional unstable manifold of $(0, -\sqrt{\delta})$ must tend to $(0, \sqrt{\delta})$, which gives the heteroclinic orbit.

To complete the proof, we have to show $\omega(z_0, -\sqrt{\delta}) = \{(0, \sqrt{\delta})\}$. Let S_0, S_1 and S_2 be defined by

$$S_0 = \{(z,v) \mid z \ge 0, v \ge 0\},$$
$$S_1 = \{(z,v) \mid z \ge 0, v + mz > 0, v < 0\},$$
$$S_2 = \left\{(z,v) \mid z \ge v(v^2 - \delta), v + mz \le 0, -\sqrt{\delta} \le v \le 0\right\}.$$

The set S_0 and the union $S = S_0 \cup S_1 \cup S_2$ are positively invariant, as can be seen by observing the vector field on their boundary. Let $(z(t), v(t))$ be the solution of

(4.6) with $(z(0), v(0)) = (z_0, -\sqrt{\delta})$. As long as the solution $(z(t), v(t))$ lies in S_1, it satisfies the inequality

$$\dot{v} \geq \delta v + z_0.$$

The solution of the ODE

$$\dot{v} = \delta v + z_0, \quad v(0) = -\sqrt{\delta}$$

grows exponentially with time. Hence, for $\delta^{3/2} < z_0 < \sqrt{\delta}/m$, the solution $(z(t), v(t))$ arrives at $\{v + mz = 0\}$ in a finite time. As the solution is monotone in S_1 but cannot tend to the origin, the solution arrives at the z-axis in a finite time. In particular, there exist $\varepsilon > 0$ and $t_1 > 0$ such that $v(t_1) = \varepsilon$. Since we have $H(z) \leq -az$, we obtain $\dot{z} = H(z)(v + mz) \leq -a\varepsilon z$, and hence, $0 \leq z(t) \leq z(t_1)e^{-a\varepsilon(t-t_1)}$. This implies that $z(t)$ exponentially decays to zero. In addition, $v(t)$ converges to $\sqrt{\delta}$ as $t \to \infty$ because we have

$$v(\delta - v^2) \leq \dot{v} \leq v(\delta - v^2) + z.$$

To demonstrate this claim, consider the following equation defining a supersolution:

$$\dot{q} = f(q) + \gamma e^{-\alpha t},$$

where $f(q) = (\delta - q^2)q$ is a function, α and γ are positive constants. Introducing a new variable $p = e^{-\alpha t}$, we obtain

$$\dot{q} = (\delta - q^2)q + \gamma p, \quad \dot{p} = -\alpha p.$$

Let β be a constant with $\beta > \gamma^2/(4\alpha)$, and let $F(q)$ be a primitive function of f. Then the function

$$L(q, p) = -F(q) + \frac{\beta}{2}p^2$$

is a Lyapunov function of the system. If $q(0) > 0$, then $q(t)$ converges to $\sqrt{\delta}$ as $t \to \infty$. Therefore, we obtain $\omega(z_0, -\sqrt{\delta}) = \{(0, \sqrt{\delta})\}$. □

To study the general cases where $\theta_{-\infty} \in (\pi/2, \pi)$, we need to know the properties of equilibria of (4.4).

Lemma 4.7 *([15]) For any $\phi \in [0, 2\pi]$, $(z, r, \theta) = (0, \sqrt{\delta}, \phi)$ is an equilibrium of (4.4), with eigenvalues*

$$0, \quad -2\delta, \quad -c\sqrt{\delta}\cos\phi.$$

The eigenspace associated with the eigenvalue $-c\sqrt{\delta}\cos\phi$ is

$$E^u(0, \sqrt{\delta}, \phi) = \text{span}\left\{\left(1, \frac{\cos\phi}{2\delta - c\sqrt{\delta}\cos\phi}, \frac{\tan\phi}{c\delta}\right)\right\}.$$

The equilibria have a zero eigenvalue because they form a one-parameter family. We only consider $\phi \in [0, \pi]$. Let S be defined by

$$S = \{(z, r, \theta) \mid z > 0, r > 0, \theta \in (0, \pi)\}.$$

By investigating the vector field on its boundary, we obtain the following lemma.

Lemma 4.8 *([15]) The subset* $S = \{(z, r, \theta) \mid z > 0, r > 0, \theta \in (0, \pi)\}$ *is invariant with respect to* (4.4).

The vector field of the (z, r, θ)-system is as follows. The next lemma is similar to [15, Lemma 3.5] but slightly different because we consider $m \geq 0$.

Lemma 4.9 *(i) If* $(z(0), r(0), \theta(0)) \in S$, *then* $\dot{\theta}(t) < 0$ *for any* $t > 0$.
(ii) If $z > 0$ *and* $r \cos\theta + mz < 0$, *then* $\dot{z} > 0$.
(iii) If $z > 0, \theta \in (\pi/2, \pi)$ *and* $r = \sqrt{\delta}$, *then* $\dot{r} < 0$.
(iv) If $z > 0$ *and* $r \cos\theta + mz > 0$, *then* $\dot{z} < 0$.
(v) If $z > 0, 0 < r < \sqrt{\delta}, \theta \in [0, \pi/2]$, *then* $\dot{r} > 0$.

The idea of proving the first two statements of Theorem 4.3 is to show the following two claims [15]: (1) The solution $(z(t), r(t), \theta(t))$ of (4.4) with the initial condition $(z(0), r(0), \theta(0)) = (z_0, r_0, \theta_0) \in S \cap W^u(0, \sqrt{\delta}, \theta_{-\infty})$ enters $\{0 < \theta < \pi/2\}$ in a finite time. (2) After entering $\{0 < \theta < \pi/2\}$, it converges to the equilibrium $(0, \sqrt{\delta}, \theta_\infty)$ for some θ_∞.

As we have seen in Lemma 4.9(i), $\theta(t)$ is strictly monotone decreasing with respect to $t > 0$. In addition, by Lemma 4.9(ii) and (iv), we have

$$\dot{\theta}(t) \leq -\frac{z_0}{\sqrt{\delta}} \sin\theta_0$$

as long as $\theta(t) \in (\pi/2, \pi)$. Therefore, the solution arrives at the set $\{r \cos\theta + mz = 0\}$ in a finite time. If $m = 0$, then this set coincides with $\{\theta = \pi/2\}$. For $m > 0$, it turns out that the solution arrives at $\{\theta = \pi/2\}$ in the following way. Along the solution orbit, $\dot{x} > 0$ and $\dot{v} > 0$ if $r \cos\theta + mz = v + mh(x) > 0$ and $v = r \cos\theta < 0$. By the monotonicity of $x(t)$, $v(t)$, and $w(t)$, there are just two possibilities:

1. $v(t)$ is negative, and $x(t) \to \infty$ and $v(t) \to -0$ as $t \to \infty$, or
2. there exists $t_c > 0$ such that $v(t_c) = 0$, that is, $\theta(t_c) = \pi/2$.

If the former is the case, then $\dot{v}(t)/\dot{x}(t) \to 0$ as $t \to \infty$. However, that is impossible because we have

$$\begin{aligned}
\frac{\dot{v}}{\dot{x}} &= \frac{v(\delta - r^2) + z}{v + mz} \\
&= \frac{v(\delta - r^2) - m^{-1}v + m^{-1}v + z}{v + mz} \\
&= \frac{1}{m} + \frac{v(\delta - m^{-1} - r^2)}{v + mz} > \frac{1}{m}.
\end{aligned}$$

Therefore, there exists $t_c > 0$ such that $\theta(t_c) = \pi/2$, that is, $v(t_c) = 0$. Let $(r_c, z_c) = (r(t_c), z(t_c))$. By Lemma 4.9(i), (iii), (v), the solution $(z(t), r(t), \theta(t))$ must satisfy

$$0 < \theta(t) < \pi/2, \quad r(t) > r_c, \quad z(t) < z_c$$

for any $t > t_c$, which implies that the sign of $\dot{x}(t)$ changes from negative to positive just once.

As $\dot{\theta}(t_c) = -z_c/r_c$ is negative, for a sufficiently small $\varepsilon > 0$ there exists t_1 such that $t_1 > t_c$ and $\theta(t_1) = \pi/2 - \varepsilon$. We have $\dot{z} \leq -(ar_c \cos(\pi/2 - \varepsilon))z$ for $t > t_1$, which implies that $z(t)$ exponentially decays to zero as $t \to \infty$. In addition, since

$$(\delta - r^2)r \leq \dot{r} \leq (\delta - r^2)r + z$$

holds, $r(t) \to \sqrt{\delta}$ as $t \to \infty$. Therefore, there exists $\theta_\infty \in [0, \pi/2)$ such that

$$\lim_{t\to\infty} (z(t), r(t), \theta(t)) = (0, \sqrt{\delta}, \theta_\infty).$$

Thus, $(z(t), r(t), \theta(t))$ is a heteroclinic solution from $(0, \sqrt{\delta}, \theta_{-\infty})$ to $(0, \sqrt{\delta}, \theta_\infty)$.

Assume that $m = 0$ holds. It remains to prove the inequality $\sin\theta_{-\infty} < \sin\theta_\infty$.

The function $z(t)$ is monotone increasing for $t > t_c$ and decreasing $t < t_c$. Regarding z as an independent variable in each interval, we obtain

$$\frac{d\theta}{dz} = -\frac{z}{H(z)(r(z))^2}\tan\theta,$$

where $r(z)$ is a function of z and denoted by

$$r(z) = \begin{cases} r_-(z) & (t < t_c), \\ r_+(z) & (t > t_c). \end{cases}$$

As the ODE above is separable, we obtain

$$\int_{\theta_{-\infty}}^{\frac{\pi}{2}} \frac{d\theta}{\tan\theta} = -\int_0^{z_c} \frac{zdz}{H(z)(r_-(z))^2},$$

$$\int_{\theta_\infty}^{\frac{\pi}{2}} \frac{d\theta}{\tan\theta} = -\int_0^{z_c} \frac{zdz}{H(z)(r_+(z))^2}.$$

Since $(\log \sin \theta)' = \cot \theta$ holds, $\sin \theta_{-\infty}$ and $\sin \theta_{\infty}$ are given by

$$\sin \theta_{-\infty} = \exp\left(\int_0^{z_c} \frac{z dz}{H(z)(r_-(z))^2} \right),$$

$$\sin \theta_{\infty} = \exp\left(\int_0^{z_c} \frac{z dz}{H(z)(r_+(z))^2} \right),$$

respectively.

Assuming that $\sin \theta_{-\infty} \geq \sin \theta_{\infty}$ holds, we obtain

$$\int_0^{z_c} \frac{z}{-H(z)} \left(\frac{1}{(r_-(z))^2} - \frac{1}{(r_+(z))^2} \right) dz \leq 0. \tag{4.7}$$

The integrand in the left-hand-side is positive in a small neighborhood of $z = z_c$. It is necessary for (4.7) that the integrand is negative in a certain interval. Therefore, there exists a $z_* \in (0, z_c)$ such that

$$r_+(z_*) - r_-(z_*) = 0$$

and

$$\left. \frac{d}{dz} (r_+(z) - r_-(z)) \right|_{z=z_*} > 0. \tag{4.8}$$

Now let $r_* = r_+(z_*) = r_-(z_*)$. Since we have $H(z_*) < 0$, $r_* < \sqrt{\delta}$ and $\theta_-(z_*) > \pi/2 > \theta_+(z_*)$, we must have

$$\begin{aligned} \left. \frac{d}{dz} (r_+(z) - r_-(z)) \right|_{z=z_*} &= \frac{\delta - r_*^2}{-H(z_*)} \left(-\frac{1}{\cos \theta_+(z_*)} + \frac{1}{\cos \theta_-(z_*)} \right) \\ &< 0, \end{aligned}$$

which contradicts (4.8). Thus $\sin \theta_{-\infty} < \sin \theta_{\infty}$ must hold.

Remark 4.10 If $m > 0$, then $\frac{d\theta}{dz}$ is no longer separable. Another approach is needed for proving the inequality.

Remark 4.11 One may think that Lemma 4.5 should prove the existence of a homoclinic solution to $(0, \sqrt{\delta}, \pi/2)$. However, no such solution can exist. As $\{z = 0\}$ is an invariant manifold composed of stable manifolds of equilibria, such a solution would have to satisfy $z(t) > 0$ for all $t \in \mathbb{R}$. That would imply that $\theta(t)$ is strictly monotone decreasing. It is necessary that $\theta(t) \in (\pi/2, \pi)$ for all $t \in \mathbb{R}$. Then $z(t)$ would be strictly monotone increasing, and it would be unable to approach zero.

Remark 4.12 We define the angles of incidence and reflection by heteroclinic orbits which connect two points of a curve of equilibria. Since the vector field is C^1, the heteroclinic orbits, and hence, the angle relation F, are at least continuous with respect to the angle of incidence.

Theorem 4.3 implies that there exists a function $\theta_{\text{ref}} = F(\theta_{\text{inc}})$ for (4.2), where F is parametrized by δ, m_0, m_1, m_2, and a. In particular, we have proved that $F(\theta_{\text{inc}}) \geq \theta_{\text{inc}}$ holds if $m_0 = 0$. Numerical simulation suggests that this inequality holds even for $m_0 > 0$. See Fig. 4.2. It seems that the function F satisfies the following properties [15, Hypothesis 1]:

1. $F(0) = 0$ and $F(\pi/2) = \pi/2$.
2. $F(\theta) > \theta$ for $0 < \theta < \pi/2$.
3. F is strictly monotone increasing.
4. F is at least C^1 on $[0, \pi/2]$.
5. $\lim_{\theta \to \pi/2} F'(\theta) = 1$.

4.2 Rectangular Domain: Model and Simulation

In the previous section, we studied reflection behavior of a particle on the half plane with a single wall $\{x = 0\}$. Recall the observations in Sect. 3.2: if a particle is confined to a square, it seems to approach a square-shaped periodic orbit. Does a particle model show this behavior?

Portions of this section are adapted from the author's previous work [50].

Let us begin by constructing a particle model on a rectangle $(-cL, cL) \times (-L, L)$, where $L > 0$ and $c \geq 1$. Let $\boldsymbol{q} = (x, y)$ and $\boldsymbol{p} = (v, w)$ be the position

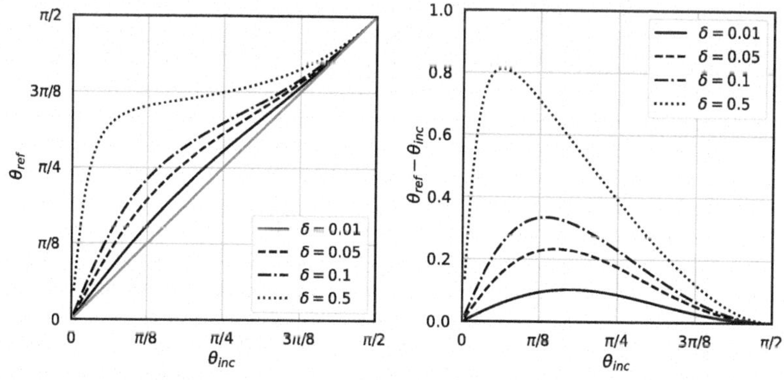

(a) $F(\theta_{inc})$ (b) $F(\theta_{inc}) - \theta_{inc}$

Fig. 4.2 The function $\theta_{\text{ref}} = F(\theta_{\text{inc}})$ for various values of δ with $m_0 = m_1 = m_2 = a = 1$. Figure adapted from the author's previous work [50] and velocity of a particle in the rectangle, respectively. Consider four mirror images of this particle with respect to each edge of the rectangle. For $j = 0, 1, 2, 3$, let $\boldsymbol{q}_j$ and $\boldsymbol{p}_j$ be a mirror image of $\boldsymbol{q}$ and its velocity, respectively. Let $d_j = |\boldsymbol{q}_j - \boldsymbol{q}|$ for $j = 0, 1, 2, 3$. The particle $\boldsymbol{q}$ is governed by the following equations:

$$\begin{cases} \dot{\boldsymbol{q}} = \boldsymbol{p} - m_0 \sum_{j=0}^{3} d_j^{-3/2} e^{-ad_j} (\boldsymbol{q}_j - \boldsymbol{q}), \\ \dot{\boldsymbol{p}} = (\delta - |\boldsymbol{p}|^2)\boldsymbol{p} - m_2 \sum_{j=0}^{3} d_j^{-3/2} e^{-ad_j} (\boldsymbol{q}_j - \boldsymbol{q}). \end{cases} \tag{4.9}$$

By the assumption that the $\boldsymbol{q}_j$'s are mirror images, the system is reduced to equations for the actual particle alone:

$$\begin{cases} \dot{x} = v + m_0(h(cL + x) - h(cL - x)), \\ \dot{v} = \left[\delta - (v^2 + w^2)\right] v + m_2(h(cL + x) - h(cL - x)), \\ \dot{y} = w + m_0(h(L + y) - h(L - y)), \\ \dot{w} = \left[\delta - (v^2 + w^2)\right] w + m_2(h(L + y) - h(L - y)), \end{cases} \tag{4.10}$$

where h is a function defined by

$$h(x) = (2x)^{-1/2} \exp(-2ax).$$

In this section, a numerical study for (4.10) is presented. The parameters are

$$\delta = 0.06, \quad L = 10.0, \quad m_0 = m_2 = a = 1$$

unless otherwise indicated.

First, let us observe trajectories of the particle on a square domain. Let $c = 1$. Figure 4.3 shows some trajectories of (4.10). The initial conditions are

$$(x, v, y, w) = (0, \sqrt{\delta}\cos\theta, 0, \sqrt{\delta}\sin\theta)$$

for $\theta = \pi, \pi + 10^{-4}, 5\pi/4$, and $13\pi/12$. When $\theta = \pi$, $(y(t), w(t)) = (0, 0)$ for all $t \geq 0$ and the particle oscillates along the x-axis. Thus, a periodic orbit is observed. On the other hand, when $\theta = \pi + 10^{-4}$, the particle eventually moves away from the periodic orbit, and it converges to another periodic orbit that is square-shaped. When $\theta = 5\pi/4$, the particle oscillates along a diagonal line but it eventually moves away from the diagonal and converges to a square-shaped cycle. Finally, when $\theta = 13\pi/12$, it converges to a square-shaped cycle. Thus, numerical simulation suggests that there are some limit cycles.

Indeed, one can locate the cycles by applying the Newton method and examine their stability. The square-shaped one is stable, while horizontal or diagonal oscil-

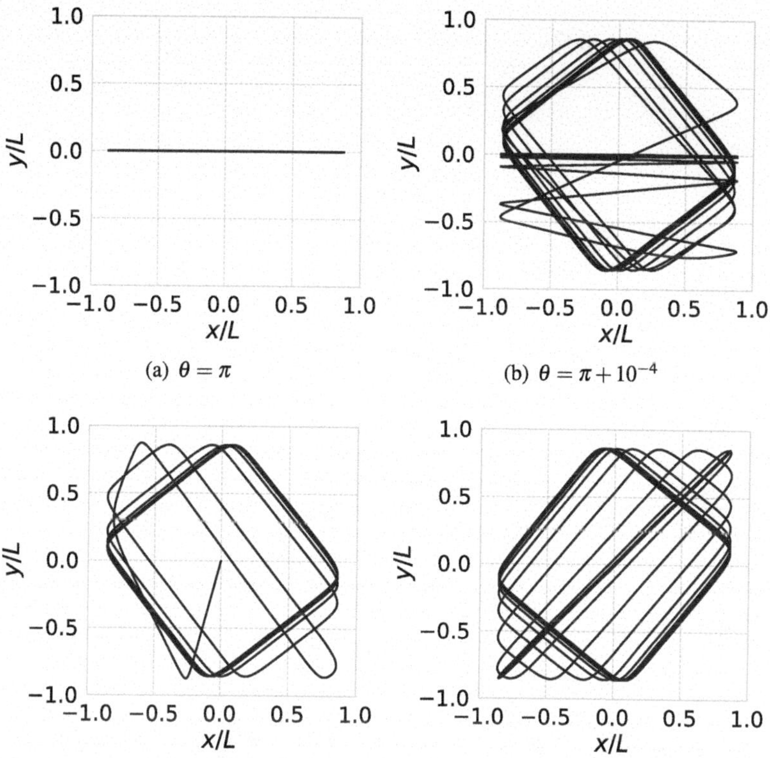

(a) $\theta = \pi$

(b) $\theta = \pi + 10^{-4}$

(c) $\theta = 5\pi/4$

(d) $\theta = 13\pi/12$

Fig. 4.3 Some trajectories in a square domain. The initial condition is $(x, v, y, w) = (0, \sqrt{\delta}\cos\theta, 0, \sqrt{\delta}\sin\theta)$ for $\theta = \pi, \pi + 10^{-4}, 5\pi/4$, and $13\pi/12$. Figure adapted from the author's previous work [49]

lations are unstable. Due to symmetry, there also exist vertical or other diagonal oscillations and square-shaped cycles of clockwise rotation. See Fig. 4.4.

Next, let us consider attractors in a rectangular domain for various values of the aspect ratio c. As a hyperbolic limit cycle is robust against perturbation, the cycles persist for c slightly greater than 1. Indeed, Fig. 4.5 shows attractors for $c = 1.1$ and 1.2, where the domain is rectangular but close to a square. We can observe a rectangular-shaped limit cycle. As c increases, the orbit tends to a diagonal line. An interesting phenomenon emerges for greater values of c. Figure 4.6 presents a trajectory for $c = 1.3$. In this case, the trajectory fills up the rectangle entirely.

In order to observe how an attractor changes depending on the aspect ratio, we draw an orbit diagram in the following way. Take $N + 1$ points $\{c_n\}_{n=0}^{N}$ on the interval $[1, 3]$, where $c_0 = 1$ and $c_N = 3$. For each $c = c_n$, iterate the Poincaré map on the hyperplane

$$\Sigma = \{(x, v, y, w) \mid y = 0, \dot{y} > 0\}$$

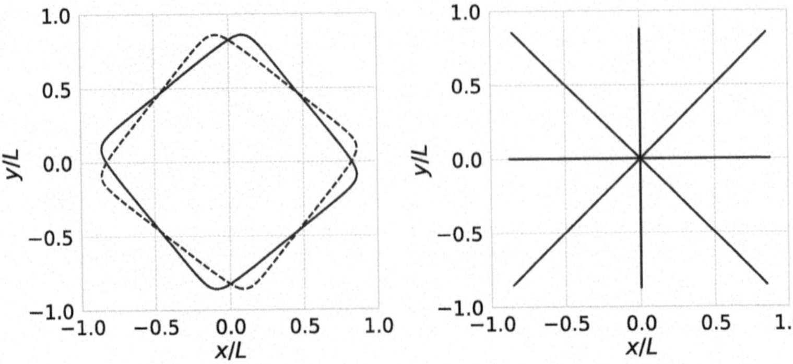

(a) Stable cycles (b) Unstable cycles

Fig. 4.4 Limit cycles of (4.9). **a** Stable cycles. The solid curve is anticlockwise, while the dashed one is clockwise. **b** Unstable cycles

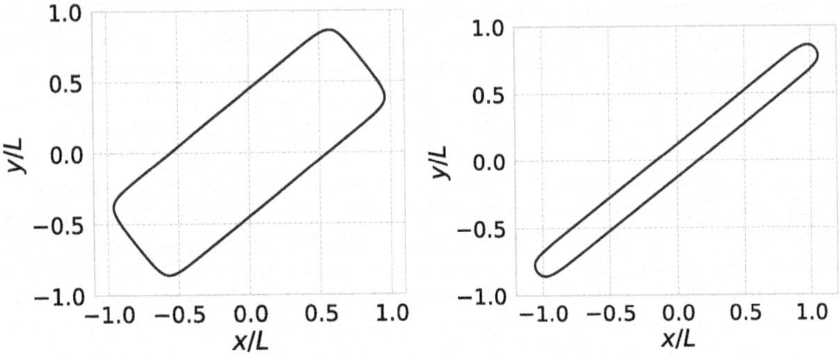

(a) $c = 1.1$ (b) $c = 1.2$

Fig. 4.5 Limit cycles on rectangular domain for $c = 1.1$ and $c = 1.2$. The initial value is $(x, v, y, w) = (0, \cos(2\pi/3), 0, \sin(2\pi/3))$. Figure adapted from the author's previous work [49]

Fig. 4.6 Trajectory of $(x, v, y, w) = (0, \cos(2\pi/3), 0, \sin(2\pi/3))$ for $20{,}000 \le t \le 80{,}000$. Figure adapted from the author's previous work [49]

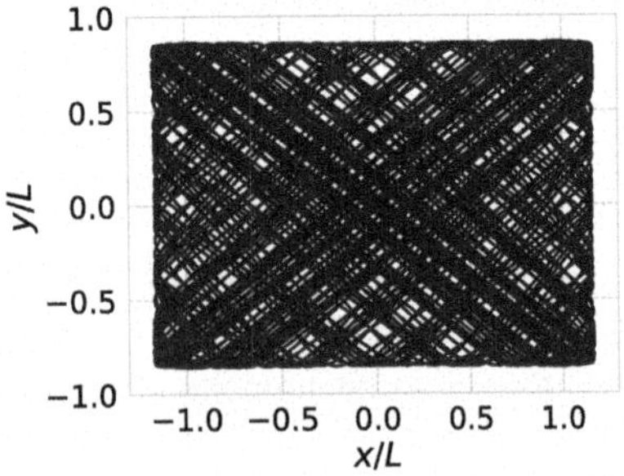

(a) $c = 1.3$

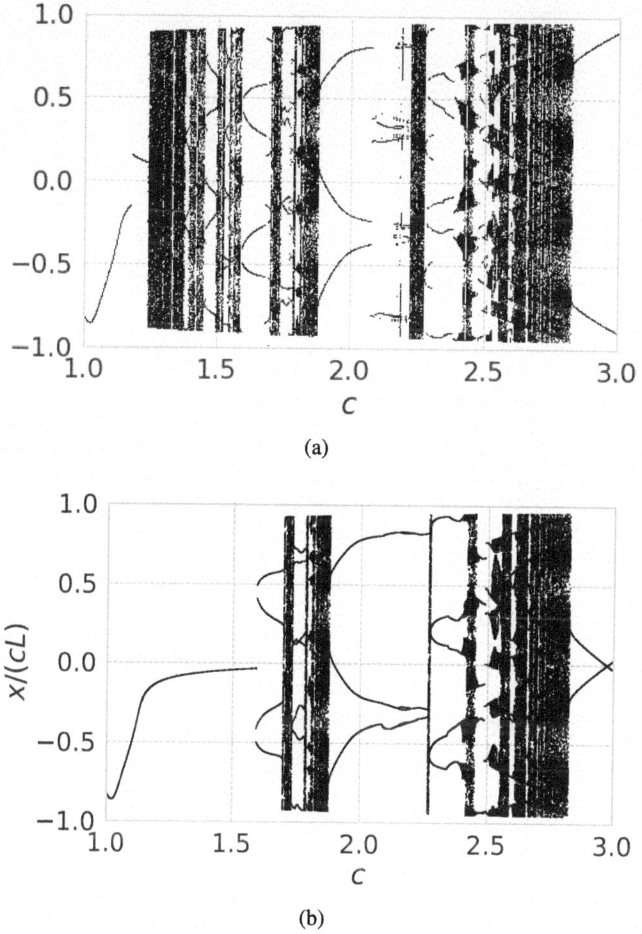

Fig. 4.7 Orbit diagram. **a** The initial point is $(x, v, w) = (0, 0.25, 0.5)$ for all c. **b** The initial point for $c = c_n$ is the last point of $c = c_{n-1}$ for each n

M times, and plot $x/(cL)$ for the last K points. An orbit diagram is presented in Fig. 4.7. The attractor may be periodic, quasiperiodic or chaotic depending on c [50]. Figures 4.8 and 4.9 show periodic orbits and nonperiodic orbits, respectively. In addition, periodic and nonperiodic attractors can coexist for the same c.

Such intermittent changes of attractors can be understood as a phase locking phenomenon. Notice that the particle model (4.10) can be regarded as a coupled oscillator system consisting of two limit cycle oscillators (x, v) and (y, w). Indeed, a periodic orbit of (4.10) corresponds to a synchronized state between the horizontal mode (x, v) and the vertical mode (y, w).

To be more precise, let us characterize the limit cycles by assigning rational numbers to them [50].

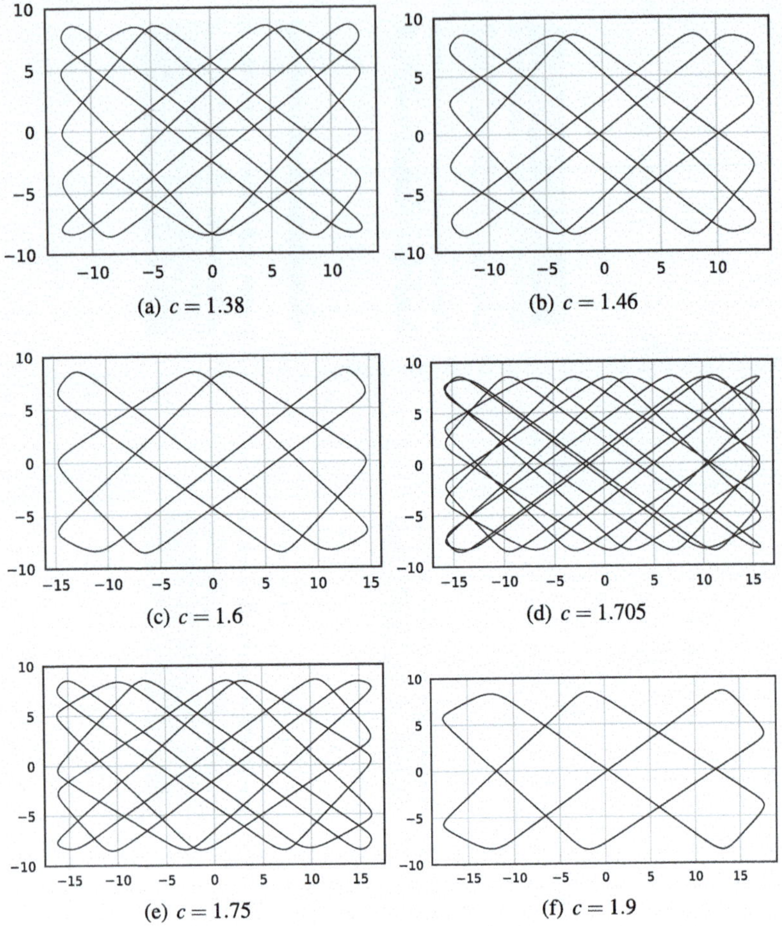

Fig. 4.8 Periodic orbits for various c. Figure adapted from the author's previous work [49]

Definition 4.13 ([50]) Suppose that, for a limit cycle of (4.10), the number of maximal points of $x(t)$ and $y(t)$ within one period are m and n, respectively. The rational number n/m is called the rotation number of the limit cycle.

For instance, rotation numbers for limit cycles in Fig. 4.8 are shown in Table 4.1. We can observe a certain structure [50] as follows. Suppose that (4.10) has limit cycles for $c = c_1$ and $c = c_2$ with $c_1 < c_2$ and the rotation numbers are ρ_1 and ρ_2, respectively. Then there is $c_0 \in (c_1, c_2)$ at which (4.10) has a limit cycle whose the rotation number is the mediant of ρ_1 and ρ_2.

Figure 4.10 presents regions of existence for cycles. The boundary of each region consists of a saddle-node bifurcation of the cycle. Figure 4.11 shows a two-parameter bifurcation diagram of the saddle-node bifurcation points on the (c, L)-plane. Each

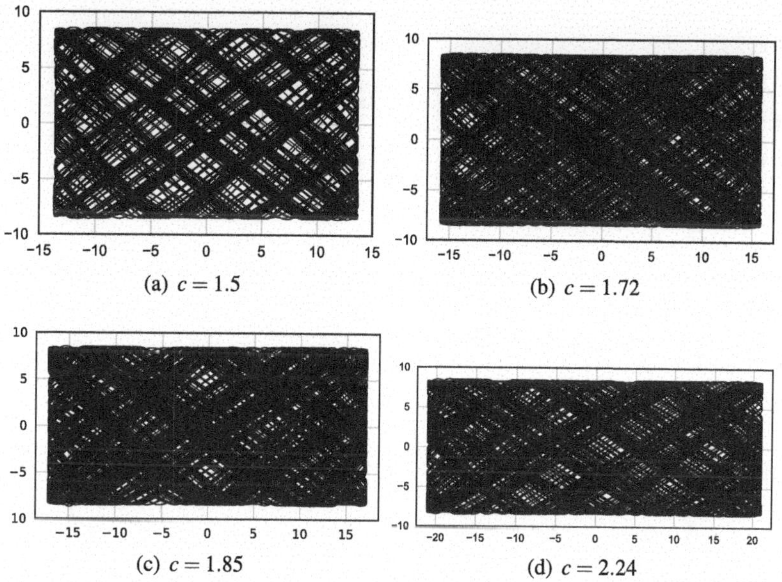

(a) $c = 1.5$ (b) $c = 1.72$

(c) $c = 1.85$ (d) $c = 2.24$

Fig. 4.9 Non-periodic orbits for various c. Figure adapted from the author's previous work [49]

Table 4.1 Rotation numbers ρ of limit cycles drawn in Fig. 4.8

	c	ρ	Comments
(a)	1.38	5/6	
(b)	1.46	4/5	4/5 = (5 + 3)/(6 + 4)
(c)	1.6	3/4	3/4 = (4 + 8)/(5 + 11) = (4 + 5)/(5 + 7) – (4 + 2)/(5 + 3)
(d)	1.705	8/11	8/11 = (3 + 5)/(4 + 7)
(e)	1.75	5/7	5/7 = (8 + 2)/(11 + 3)
(f)	1.9	2/3	

region of existence has a wedge-like shape. This suggests that the phase-locking phenomenon and Arnold tongues occur:

- There exists a family of invariant tori parameterized by two parameters.
- There are parameter regions in which the invariant torus has a rational rotation number and hyperbolic limit cycles exist on the torus (phase-locking).
- The boundary of such regions forms a wedge-like or tongue-like shape (Arnold tongues).

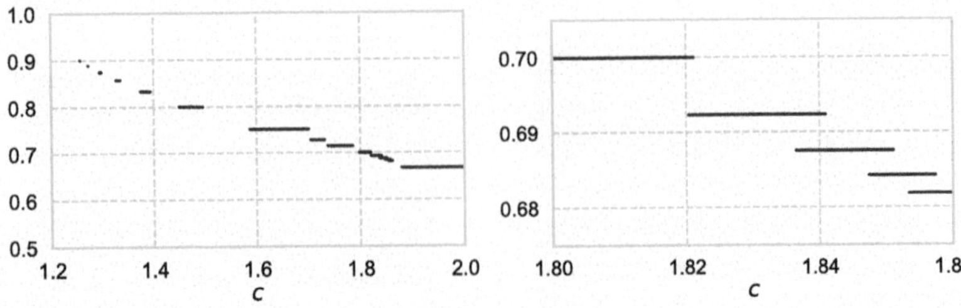

Fig. 4.10 Rotation numbers of limit cycles with respect to c. Figure adapted from the author's previous work [50, Fig. 10]

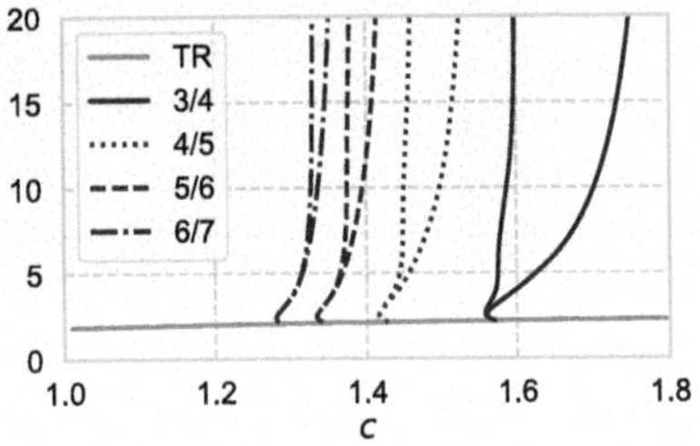

Fig. 4.11 Saddle-node bifurcation and torus bifurcation curves on the (c, L)-plane. Figure adapted from the author's previous work [50, Fig. 11]

As shown in Fig. 4.11, there exist parameter regions in which the invariant torus supports only a finite number of hyperbolic limit cycles with the same rotation number. These regions are called Arnold tongues. At the boundary of the Arnold tongues, the limit cycles undergo fold bifurcations. As L decreases, the tongues approach the curve labeled 'TR', which is a torus bifurcation curve of the horizontal oscillation. It is surprising that this complex behavior originates not from the stable square-shaped cycle but the unstable horizontal oscillation.

Figure 4.7b suggests that the rectangular-shaped cycle coexists with other attractors. In order to study its existence region, we apply a path-continuation method using the AUTO software [12]. See Fig. 4.12. The rectangular-shaped attractor loses its stability at $c \approx 1.592$ by a torus bifurcation, undergoes a saddle-node bifurcation at $c \approx 1.630$, and comes back to $c = 1.0$ with the cycle on the diagonal line. Figure 4.13 shows the Poincaré map for $L = 10, c = 1.3$ projected onto the $(x/cL, v)$-plane. If a quasiperiodic orbit bifurcates from a rectangular-shaped cycle, it should remain in a neighborhood of the cycle. However, the invariant circle corresponding to a quasiperiodic orbit is too large. Thus, the instability of rectangular-shaped cycles does not explain the intermittent change of attractors. In the next subsection, we study the organizing center of the phase locking phenomenon.

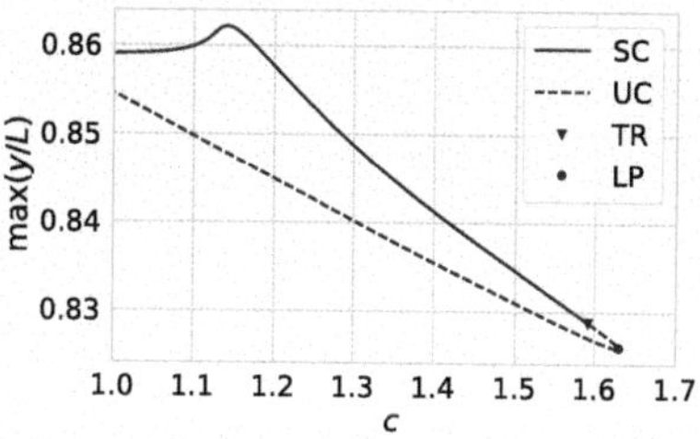

Fig. 4.12 Bifurcation diagram of the rectangular-shaped cycle. *SC* stable cycle, *UC* unstable cycle, *TR* torus bifurcation point, *LP* limit point

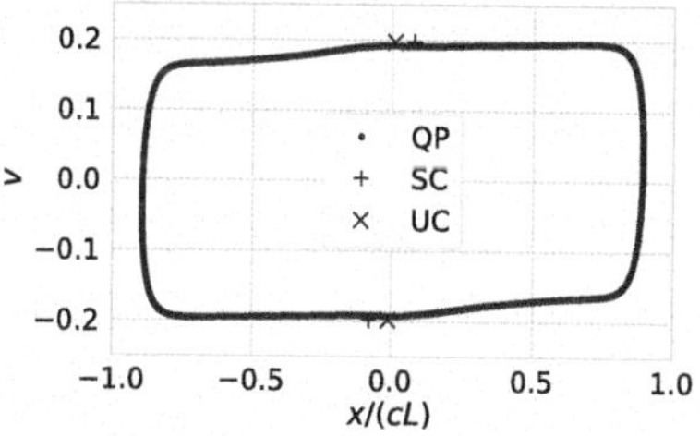

Fig. 4.13 Poincaré section Σ. *QP* quasiperiodic orbit, *SC* stable cycle, *UC* unstable cycle

4.3 Square Domain: Mathematical Analysis

In this section, we assume $c = 1$, to study the particle model in a square domain, which is given by (4.10) with $L > 0$. It is easy to see that the origin $(x, v, y, w) = (0, 0, 0, 0)$ is an equilibrium. We consider a bifurcation of the origin under the variation of δ. We assume $\delta < m_2/m_0$. If $\delta > m_2/m_0$, then nontrivial equilibria exist. We would like to exclude such a case.

Theorem 4.14 *Assume that* $c = 1, L > 0$ *and* $\delta < m_2/m_0$, *and consider* (4.11). *The origin* $(x, v, y, w) = (0, 0, 0, 0)$ *is an equilibrium for any* δ. *It is linearly stable for* $\delta < -2m_0h'(L)$, *while it is unstable for* $\delta > -2m_0h'(L)$. *Moreover, five branches of limit cycles exist in at least a small neighborhood of* $(x, v, y, w, \delta) = (0, 0, 0, 0, -2m_0h'(L))$ *with* $\delta > -2m_0h'(L)$.

1. *Limit cycles with* $y = w = 0$ *or those with* $x = v = 0$ *are unstable.*
2. *Limit cycles with* $x = y, v = w$ *or* $x = -y, v = -w$ *are unstable.*
3. *Any other limit cycle is stable.*

Let $\eta = 2m_2h'(L)$ and $m_0 = mm_2$. Since $c = 1$, the linearized matrix of (4.10) at the origin is given by

$$J = \begin{pmatrix} J_1 & O_2 \\ O_2 & J_1 \end{pmatrix}, \quad J_1 = \begin{pmatrix} m\eta & 1 \\ \eta & \delta \end{pmatrix},$$

where O_2 is the 2×2 zero matrix. Obviously, all eigenvalues have algebraic multiplicity 2. The origin is stable for $\delta < \ m\eta = \ 2m_0h'(L)$, but unstable for $\delta > -2m_0h'(L)$. The matrix has just two purely imaginary eigenvalues $\pm i\sqrt{-m^2\eta^2 - \eta}$

at $\delta = -m\eta$. Here, we assume $-m^2\eta < 1$. As δ varies, these eigenvalues cross the imaginary axis with nonzero speed.

Suppose that $\lambda = \sigma + i\omega$ is an eigenvalue of J_1. Let $\boldsymbol{q} \in \mathbb{C}^2$ and $\boldsymbol{p} \in \mathbb{C}^2$ be defined by

$$\boldsymbol{q} = \begin{pmatrix} 1 \\ -m\eta + \lambda \end{pmatrix}, \quad \boldsymbol{p} = \frac{i}{2\omega} \begin{pmatrix} \overline{\lambda} - \delta \\ 1 \end{pmatrix},$$

and let $\boldsymbol{q}_1, \boldsymbol{q}_2, \boldsymbol{p}_1, \boldsymbol{p}_2 \in \mathbb{C}^4$ be

$$\boldsymbol{q}_1 = \begin{pmatrix} \boldsymbol{q} \\ \boldsymbol{0} \end{pmatrix}, \boldsymbol{q}_2 = \begin{pmatrix} \boldsymbol{0} \\ \boldsymbol{q} \end{pmatrix}, \boldsymbol{p}_1 = \begin{pmatrix} \boldsymbol{p} \\ \boldsymbol{0} \end{pmatrix}, \boldsymbol{p}_2 = \begin{pmatrix} \boldsymbol{0} \\ \boldsymbol{p} \end{pmatrix},$$

where $\boldsymbol{0} \in \mathbb{C}^2$. Then we have

$$J\boldsymbol{q}_j = \lambda \boldsymbol{q}_j, \; J^T \boldsymbol{p}_j = \overline{\lambda} \boldsymbol{p}_j$$

for $j = 1, 2$, and $\langle \boldsymbol{p}_j, \boldsymbol{q}_k \rangle = \delta_{jk}$ for $j, k = 1, 2$.

Notice that the system has a symmetry. Let R_1, R_2, and S be given by

$$R_1 = \begin{pmatrix} -I_2 & O \\ O & I_2 \end{pmatrix}, \quad R_2 = \begin{pmatrix} I_2 & O \\ O & -I_2 \end{pmatrix}, \quad S = \begin{pmatrix} O & I_2 \\ I_2 & O \end{pmatrix}.$$

Let Γ be the group generated by $\{R_1, R_2, S\}$, that is,

$$\Gamma = \{I_4, -I_4, R_1, R_2, S, SR_1, SR_2, -S\},$$

where the use of the symbol Γ is different from the previous chapter. Γ gives a representation of actions of the dihedral group $\mathbf{D}_4$ on $\mathbb{R}^4$. If $c = 1$, then (4.10) commutes with Γ. Hence, if $(x(t), v(t), y(t), w(t))$ is a solution of (4.10), then $\gamma(x(t), v(t), y(t), w(t))$ is also a solution for any $\gamma \in \Gamma$.

The Taylor expansion of the vector field up to third order terms is given by

$$\dot{\boldsymbol{u}} = J\boldsymbol{u} + C(\boldsymbol{u}) + O(|\boldsymbol{u}|^5),$$

where $\boldsymbol{u} = (x, v, y, w)$, and $C(\boldsymbol{u})$ is defined by

$$C(\boldsymbol{u}) = \frac{1}{6} \begin{pmatrix} m\xi x^3 \\ \xi x^3 - 6v^3 - 6vw^2 \\ m\xi y \\ \xi y^3 - 6v^2 w - 6w^3 \end{pmatrix}.$$

One can rewrite the vector field in new coordinates $(z_1, z_2) \in \mathbb{C}^2$ as

$$\boldsymbol{u} = z_1 \boldsymbol{q}_1 + z_2 \boldsymbol{q}_2 + \overline{z}_1 \overline{\boldsymbol{q}}_1 + \overline{z}_2 \overline{\boldsymbol{q}}_2.$$

Since the vector field commutes with Γ, it can be transformed into the following normal form around $(z_1, z_2) = (0, 0)$:

$$\begin{cases} \dot{z}_1 = \lambda z_1 + c_1 z_1 |z_1|^2 + c_2 z_1 |z_2|^2 + c_3 \bar{z}_1 z_2^2 + \cdots, \\ \dot{z}_2 = \lambda z_2 + c_2 z_1 |z_1|^2 + c_1 z_1 |z_2|^2 + c_3 z_1^2 \bar{z}_2 + \cdots, \end{cases}$$

where $\lambda = \sigma + \mathrm{i}\omega$ and c_1, c_2, and c_3 are given by

$$c_1 = \frac{m\xi + 6\eta}{4} + \mathrm{i}\frac{6m\eta^2 - (m^2\eta + 1)\xi}{4\omega_0},$$

$$c_2 = \eta + \mathrm{i}\frac{m\eta^2}{\omega_0}, \quad c_3 = \frac{\eta}{2} + \mathrm{i}\frac{m\eta^2}{2\omega_0},$$

and $\xi = 2m_2 h'''(L)$. Let $a_j = \mathrm{Re}(c_j)$ and $b_j = \mathrm{Im}(c_j)$ for $j = 1, 2, 3$. Truncating the right hand side up to third order terms and introducing polar coordinates $z_j = r_j e^{\mathrm{i}\theta_j}$ $(j = 1, 2)$, we obtain

$$\begin{cases} \dot{r}_1 = r_1 \left(\sigma + a_1 r_1^2 + a_2 r_2^2\right) + r_1 r_2^2 \mathrm{Re}(c_3 e^{-2\mathrm{i}\phi}), \\ \dot{r}_2 = r_2 \left(\sigma + a_2 r_1^2 + a_1 r_2^2\right) + r_1^2 r_2 \mathrm{Re}(c_3 e^{2\mathrm{i}\phi}), \\ \dot{\theta}_1 = \omega + b_1 r_1^2 + b_2 r_2^2 + r_2^2 \mathrm{Im}(c_3 e^{-2\mathrm{i}\phi}), \\ \dot{\theta}_2 = \omega + b_2 r_1^2 + b_1 r_2^2 + r_1^2 \mathrm{Im}(c_3 e^{2\mathrm{i}\phi}), \end{cases}$$

where $\phi = \theta_1 - \theta_2$. Note that ϕ satisfies

$$\dot{\phi} = (b_1 - b_2)(r_1^2 - r_2^2) + r_2^2 \mathrm{Im}(c_3 e^{-2\mathrm{i}\phi}) - r_1^2 \mathrm{Im}(c_3 e^{2\mathrm{i}\phi}).$$

Thus, we only have to consider the equations for (r_1, r_2, ϕ).

The nontrivial equilibria are as follows:

- $(r_1, r_2, \phi) = (\sqrt{-\sigma/a_1}, 0, 0)$,
- $(r_1, r_2, \phi) = (0, \sqrt{-\sigma/a_1}, 0)$,
- $(r_1, r_2, \phi) = (\sqrt{-\sigma/(a_1 + a_2 + a_3)}, \sqrt{-\sigma/(a_1 + a_2 + a_3)}, 0)$, and
- $(r_1, r_2, \phi) = (\sqrt{-\sigma/(a_1 + a_2 - b_3)}, \sqrt{-\sigma/(a_1 + a_2 - b_3)}, \pi/2)$,

which correspond to horizontal, vertical and diagonal oscillations, and the square-shaped cycle, respectively. All of them bifurcate for $\sigma > 0$ because $a_1, a_2, a_3, a_1 + a_2 + a_3$ and $a_1 + a_2 - b_3$ are negative.

The stability of these cycles is evaluated in the following way. See [77] for details. The truncated normal form is written as

$$\begin{cases} \dot{z}_1 = \lambda z_1 + \alpha_u z_2 (z_1 \bar{z}_2 + \bar{z}_1 z_2) + \alpha_v z_2 (z_1 \bar{z}_2 - \bar{z}_1 z_2) + \alpha_w z_1 (|z_1|^2 - |z_2|^2), \\ \dot{z}_2 = \lambda z_2 + \alpha_u z_1 (z_1 \bar{z}_2 + \bar{z}_1 z_2) - \alpha_v z_1 (z_1 \bar{z}_2 - \bar{z}_1 z_2) - \alpha_w z_2 (|z_1|^2 - |z_2|^2), \end{cases}$$

where $\alpha_u, \alpha_v, \alpha_w$ are given by

$$\begin{cases} \alpha_u = R_u + \mathrm{i}I_u = (c_1 + c_2 + c_3)/2, \\ \alpha_v = R_v + \mathrm{i}I_v = (c_1 + c_2 - c_3)/2, \\ \alpha_w = R_w + \mathrm{i}I_w = c_1. \end{cases}$$

Introducing new variables[1] $u, v, w, \psi \in \mathbb{R}$ by

$$u + \mathrm{i}v = 2z_1\bar{z}_2, \quad w = |z_1|^2 - |z_2|^2, \quad e^{\mathrm{i}\psi} = \frac{z_1 z_2}{|z_1 z_2|},$$

we obtain

$$\begin{cases} \frac{1}{2}\dot{u} = u(\sigma + R_u r) + (I_v - I_w)vw, \\ \frac{1}{2}\dot{v} = v(\sigma + R_v r) + (I_w - I_u)wu, \\ \frac{1}{2}\dot{w} = w(\sigma + R_w r) + (I_u - I_v)uv, \\ \frac{1}{2}\dot{\psi} = \omega + O(r), \end{cases}$$

where $r = \sqrt{u^2 + v^2 + w^2}$. Since u, v, w are independent of ψ, we only have to consider the (u, v, w)-system. The horizontal oscillation, the diagonal oscillation, and square-shaped cycle correspond to the following equilibria:

- w-solution: $w \neq 0, u = v = 0, \sigma + R_w r = 0$,
- u-solution: $u \neq 0, v = w = 0, \sigma + R_u r = 0$,
- v-solution: $v \neq 0, w = u = 0, \sigma + R_v r = 0$.

The stability of each cycle is determined by that of the corresponding equilibrium. First, let us consider the w-solution. The linearized matrix around the w-solution is

$$2\begin{pmatrix} \sigma + R_u r & (I_v - I_w)w & 0 \\ (I_w - I_u)w & \sigma + R_v r & 0 \\ 0 & 0 & -\sigma \end{pmatrix}.$$

It has -2σ as an eigenvalue. The other two are determined by the 2×2 matrix $2|w|J_s$ with J_s satisfying

$$\frac{1}{2}\mathrm{tr}\, J_s = (R_u + R_v - 2R_w)r = (a_2 - a_1)r > 0.$$

Therefore, the w-solution, and hence horizontal or vertical oscillation, is unstable because it has at least one eigenvalue with a positive real part.

[1] Here, v and w are different from those of the particle model.

Next, the linearized matrix around the u-solution is

$$2\begin{pmatrix} -\sigma & 0 & 0 \\ 0 & \sigma + R_v r & (I_w - I_u)u \\ 0 & (I_u - I_v)u & \sigma + R_w r \end{pmatrix}.$$

This matrix also has an eigenvalue -2σ. The other two eigenvalues are those of a matrix J_s with

$$\begin{aligned} \operatorname{tr} J_s &= r(a_1 - a_2 - 3a_3) = \frac{r}{2}(m\xi - 4\eta), \\ \det J_s &= \frac{r^2}{2}(a_3(a_3 + a_2 - a_1) + b_3(b_3 + b_2 - b_1)) \\ &= -\frac{r^2}{2}\frac{m\eta\xi}{4} \leq 0. \end{aligned}$$

If $m > 0$, then the u-solution is unstable.

Finally, the linearized matrix around the v-solution is

$$2\begin{pmatrix} \sigma + R_u r & 0 & (I_v - I_w)v \\ 0 & -\sigma & 0 \\ (I_u - I_v)v & 0 & \sigma + R_w r \end{pmatrix}.$$

Again, this matrix has -2σ as an eigenvalue. The other two eigenvalues are those of a matrix J_s with

$$\begin{aligned} \operatorname{tr} J_s &= r(a_1 - a_2 + 3a_3) = \frac{r}{4}(m\xi + 8\eta) < 0, \\ \det J_s &= \frac{r^2}{2}(a_3(a_1 - a_2 + a_3) + b_3(b_1 - b_2 + b_3)) \\ &= \frac{r^2}{2}\frac{\eta(-2\eta^2 + m\xi\omega^2)}{4\omega^2} > 0. \end{aligned}$$

This implies that the v-solution is stable.

In summary, the rest state of (4.10) with $c = 1$ undergoes a Hopf bifurcation, horizontal, vertical, and diagonal oscillations are unstable, and the square-shaped cycle is stable.

4.4 Rectangular Domain: Mathematical Analysis

In this section, we analyze the particle model (4.10) on a rectangular domain. We assume $c > 1$ throughout this section. We will see that the rest state $(x, v, y, w) = (0, 0, 0, 0)$ undergoes a bifurcation, which results in Arnold tongues. The bifurcation is the organizing center of intermittent changes of attractors. Portions of this section are adapted from the author's previous work [50].

Since the linearized matrix at the origin for (4.10) is given by

$$\begin{pmatrix} 2m_0h'(cL) & 1 & 0 & 0 \\ 2m_2h'(cL) & \delta & 0 & 0 \\ 0 & 0 & 2m_0h'(L) & 1 \\ 0 & 0 & 2m_2h'(L) & \delta \end{pmatrix},$$

the origin has purely imaginary eigenvalues at $\delta = -2m_0h'(cL)$ and $\delta = -2m_0h'(L)$. It is not difficult to prove that horizontal and vertical oscillations bifurcate from these points via Hopf bifurcations. However, we would like to obtain more insights according to [50]. Instead of (4.10), we investigate

$$\begin{cases} \dot{x} = v + m_0\left(h(cL+x) - h(cL-x)\right), \\ \dot{v} = \left[\delta_1 - (v^2+w^2)\right]v + m_2\left(h(cL+x) - h(cL-x)\right), \\ \dot{y} = w + m_0\left(h(L+y) - h(L-y)\right), \\ \dot{w} = \left[\delta_2 - (v^2+w^2)\right]w + m_2\left(h(L+y) - h(L-y)\right), \end{cases} \tag{4.11}$$

where δ_1 and δ_2 are real numbers. We recover the original system (4.10) if we vary the parameters along the diagonal line $\{(\delta_1, \delta_2) \mid \delta_1 = \delta_2 = \delta\}$.

In a manner similar to the previous section, define $\eta_k, \xi_k (k = 1, 2)$ by

$$\eta_1 = 2h'(cL), \quad \eta_2 = 2h'(L), \quad \xi_1 = 2h'''(cL), \quad \xi_2 = 2h'''(L).$$

Introducing new parameters $\boldsymbol{\alpha} = (\alpha_1, \alpha_2)$ by

$$\alpha_1 = \frac{1}{2}(\delta_1 + m_0\eta_1), \quad \alpha_2 = \frac{1}{2}(\delta_2 + m_0\eta_2),$$

we denote (4.11) compactly by

$$\dot{\boldsymbol{u}} = f(\boldsymbol{u}, \boldsymbol{\alpha}), \quad \boldsymbol{u} = (x, v, y, w). \tag{4.12}$$

The origin $\boldsymbol{u} = \boldsymbol{0}$ possesses two pairs of purely imaginary eigenvalues at $\boldsymbol{\alpha} = (0, 0)$. Indeed, the linearized matrix at the origin is

$$J = \begin{pmatrix} m_0\eta_1 & 1 & 0 & 0 \\ m_2\eta_1 & \alpha_1 - m_0\eta_1 & 0 & 0 \\ 0 & 0 & m_0\eta_2 & 1 \\ 0 & 0 & m_2\eta_2 & \alpha_2 - m_0\eta_2 \end{pmatrix}. \tag{4.13}$$

The eigenvalues of J are

$$\lambda_k(\alpha_k) = \alpha_k + \frac{1}{2}\sqrt{4\left[(\alpha_k - m_0\eta_k)^2 + m_2\eta_k\right]} \quad (k = 1, 2)$$

and their complex conjugates. Note that

$$\lambda_k(0) = \frac{1}{2}\sqrt{4(m_0^2\eta_k^2 + m_2\eta_k)} \quad (k = 1, 2)$$

is purely imaginary and that there exists a neighborhood U of $\boldsymbol{\alpha} = (0, 0)$ in which J has two pairs of complex eigenvalues [50, Proposition 3]. For $\boldsymbol{\alpha} \in U$, we denote

$$\omega_k(\alpha_k) = \frac{1}{2}\sqrt{-4\left[(\alpha_k - m_0\eta_k)^2 + m_2\eta_k\right]},$$

which is positive for $k = 1, 2$. The eigenvalues of J are given by

$$\lambda_k = \alpha_k + \mathrm{i}\omega_k(\alpha_k), \quad \overline{\lambda_k} = \alpha_k - \mathrm{i}\omega_k(\alpha_k) \quad (k = 1, 2).$$

Let $\boldsymbol{q}_1, \boldsymbol{q}_2, \boldsymbol{p}_1, \boldsymbol{p}_2 \in \mathbb{C}^4$ be

$$\boldsymbol{q}_1 = \begin{pmatrix} 1 \\ -m_0\eta_1 + \lambda_1 \\ 0 \\ 0 \end{pmatrix}, \quad \boldsymbol{q}_2 = \begin{pmatrix} 0 \\ 0 \\ 1 \\ -m_0\eta_2 + \lambda_2 \end{pmatrix},$$
$$\boldsymbol{p}_1 = \frac{\mathrm{i}}{2\omega_1}\begin{pmatrix} m_0\eta_1 - \lambda_1 \\ 1 \\ 0 \\ 0 \end{pmatrix}, \quad \boldsymbol{p}_2 = \frac{\mathrm{i}}{2\omega_2}\begin{pmatrix} 0 \\ 0 \\ m_0\eta_2 - \lambda_2 \\ 1 \end{pmatrix}.$$

We have

$$J\boldsymbol{q}_k = \lambda_k\boldsymbol{q}_k, \quad J^T\boldsymbol{p}_k = \overline{\lambda_k}\boldsymbol{p}_k \quad (k = 1, 2),$$
$$\langle \boldsymbol{p}_i, \boldsymbol{q}_j \rangle = \delta_{ij}, \quad \langle \boldsymbol{p}_i, \overline{\boldsymbol{q}}_j \rangle = 0, \quad (i, j = 1, 2).$$

In this case, (4.11) has the $\mathbf{D}_2$-symmetry [50, Proposition 1].

Proposition 4.1 *([50]) For any $\boldsymbol{\alpha} \in \mathbb{R}^2$,* (4.12) *commutes with the linear transformations $-I_4$, R_1 and R_2 defined by*

$$R_1 = \begin{pmatrix} -I_2 & O \\ O & I_2 \end{pmatrix}, \quad R_2 = \begin{pmatrix} I_2 & O \\ O & -I_2 \end{pmatrix},$$

where O is the zero matrix of order two and I_k is the identity matrix of order k for $k = 2, 4$. In other words, f satisfies

$$\gamma f(\boldsymbol{u}, \boldsymbol{\alpha}) = f(\gamma\boldsymbol{u}, \boldsymbol{\alpha})$$

for any $\gamma \in \{I_4, R_1.R_2, -I_4\}$ and $(\boldsymbol{u}, \boldsymbol{\alpha}) \in \mathbb{R}^4 \times \mathbb{R}^2$.

Let $\Gamma = \{I_4, R_1.R_2, -I_4\}$. The set Γ forms a group with respect to matrix multiplications, defining an action of the dihedral group $\mathbf{D}_2$ on $\mathbb{R}^4$. By the above proposition, the Taylor expansion of f at $\boldsymbol{u} = \mathbf{0}$ is given by

$$f(\boldsymbol{u}, \boldsymbol{\alpha}) = J\boldsymbol{u} + C(\boldsymbol{u}) + O(\|\boldsymbol{u}\|^5),$$

where the higher-order terms consist of only odd order terms and the function $C(\boldsymbol{u})$ is given by

$$C(\boldsymbol{u}) = \frac{1}{6}\begin{pmatrix} m_0\xi_1 x^3 \\ m_2\xi_1 x^3 - 6v(v^2 + w^2) \\ m_0\xi_2 y^3 \\ m_2\xi_2 y^3 - 6w(v^2 + w^2) \end{pmatrix}.$$

Since the system has $\mathbf{D}_2$-symmetry, we can transform it into the following normal form in a small neighborhood of $(\boldsymbol{u}, \boldsymbol{\alpha}) = (\mathbf{0}, \mathbf{0})$ ([50, Theorem 1]):

$$\begin{cases} \dot{\zeta}_1 = \zeta_1(\lambda_1 + c_{11}|\zeta_1|^2 + c_{12}|\zeta_2|^2) + O(\|\boldsymbol{\zeta}\|^4), \\ \dot{\zeta}_2 = \zeta_2(\lambda_2 + c_{21}|\zeta_1|^2 + c_{22}|\zeta_2|^2) + O(\|\boldsymbol{\zeta}\|^4), \end{cases} \tag{4.14}$$

where $c_{ij} \in \mathbb{C}(i, j = 1, 2)$. See [50] for details of calculation. To analyze the bifurcation, $\gamma_{ij} = \mathrm{Re}(c_{ij})$ are important. These are given as follows:

$$\gamma_{11} = \frac{m_0\xi_1 + 6m_1m_2\eta_1}{4}, \quad \gamma_{22} = \frac{m_0\xi_2 + 6m_1m_2\eta_2}{4},$$
$$\gamma_{12} = m_1m_2\eta_2, \quad \gamma_{21} = m_1m_2\eta_1.$$

Introducing the polar coordinates $\zeta_k = \rho_k e^{\mathrm{i}\varphi_k}$, $k = 1, 2$, we can rewrite (4.14) as

$$\begin{cases} \dot{\rho}_1 = \rho_1(\alpha_1 + \gamma_{11}\rho_1^2 + \gamma_{12}\rho_2^2) + \Phi_1, \\ \dot{\rho}_2 = \rho_2(\alpha_2 + \gamma_{21}\rho_1^2 + \gamma_{22}\rho_2^2) + \Phi_2, \\ \dot{\varphi}_1 = \omega_1 + \Psi_1, \\ \dot{\varphi}_2 = \omega_2 + \Psi_2, \end{cases}$$

where Φ_1, Φ_2, Ψ_1, and Ψ_2 are higher-order terms. We truncate the higher order terms, leaving

$$\begin{cases} \dot{\rho}_1 = \rho_1(\alpha_1 + \gamma_{11}\rho_1^2 + \gamma_{12}\rho_2^2), \\ \dot{\rho}_2 = \rho_2(\alpha_2 + \gamma_{21}\rho_1^2 + \gamma_{22}\rho_2^2), \end{cases} \tag{4.15}$$

and

$$\dot{\varphi}_1 = \omega_1, \qquad \dot{\varphi}_2 = \omega_2. \tag{4.16}$$

Note that the equations for amplitudes (ρ_1, ρ_2) and phases (φ_1, φ_2) are decoupled from each other.

Equation (4.15) is well known in bifurcation theory ([5, Sect. 4.5], [24, Sect. 7.5], [43, Sect. 8.6]). It can exhibit rich dynamics, depending on the signs of coefficients. However, since we have that $\gamma_{11} < 0$ and $\gamma_{22} < 0$, our case is classified as the *simple case* in [43, Sect. 8.6]. Indeed, (4.15) undergoes only bifurcations of equilibria.

Equation (4.15) has at most four equilibria. Their properties are summarized as follows ([50, Lemmas 7–9]). For any $\boldsymbol{\alpha}$, the origin $(\rho_1, \rho_2) = (0, 0)$ is an equilibrium. It is linearly stable if $\alpha_1 < 0$ and $\alpha_2 > 0$, while it is unstable if $\alpha_1 > 0$ or $\alpha_2 > 0$. There are two bifurcation curves of the origin:

$$H_1 = \{(\alpha_1, \alpha_2) \mid \alpha_1 = 0\}, \quad H_2 = \{(\alpha_1, \alpha_2) \mid \alpha_2 = 0\}.$$

Equilibria $E_1 = (\sqrt{-\alpha_1/\gamma_{11}}, 0)$ and $E_2 = (0, \sqrt{-\alpha_2/\gamma_{22}})$ bifurcate from H_1 and H_2, respectively. E_1 exists for $\alpha_1 > 0$, and it is linearly stable if $\alpha_1 > 0$ and $\gamma_{11}\alpha_2 > \gamma_{21}\alpha_1$. On the other hand, E_2 exists for $\alpha_2 > 0$, and it is linearly stable if $\alpha_2 > 0$ and $\gamma_{12}\alpha_2 < \gamma_{22}\alpha_1$. There are bifurcation curves T_1 and T_2

$$T_1 = \{(\alpha_1, \alpha_2) \mid \gamma_{21}\alpha_1 = \gamma_{11}\alpha_2\},$$

and

$$T_2 = \{(\alpha_1, \alpha_2) \mid \gamma_{22}\alpha_1 = \gamma_{12}\alpha_2\}.$$

When the parameter crosses these curves, E_1 or E_2 becomes unstable and the non-trivial equilibrium E_3 bifurcates. It is given by

$$E_3 = \left(\sqrt{\frac{\gamma_{12}\alpha_2 - \gamma_{22}\alpha_1}{\gamma_{11}\gamma_{22} - \gamma_{12}\gamma_{21}}}, \sqrt{\frac{\gamma_{21}\alpha_1 - \gamma_{11}\alpha_2}{\gamma_{11}\gamma_{22} - \gamma_{12}\gamma_{21}}} \right)$$

for

$$\frac{\gamma_{21}}{\gamma_{11}}\alpha_1 < \alpha_2 < \frac{\gamma_{22}}{\gamma_{12}}\alpha_1,$$

and is linearly stable.

Remark that the full system contains the phase equation. The equilibria of the amplitude equation are interpreted in the following manner [50].

- The trivial equilibrium of (4.15) corresponds to that of (4.14).
- The curves H_1 and H_2 correspond to Hopf bifurcation curves of the origin for (4.14).
- The equilibria E_1 and E_2 are limit cycles of (4.14), emanating from H_1 and H_2, respectively.
- The curves T_1 and T_2 correspond to Neimark–Sacker bifurcation curves of (4.14).
- The equilibrium E_3 corresponds to an invariant torus of (4.14).

These invariant sets and bifurcation curves persist under small perturbation by the higher order terms. Thus, we arrive at the following theorem:

Theorem 4.15 *([50]) For sufficiently small* $\|\alpha\|$, (4.12) *yields bifurcation curves* $\tilde{H}_1, \tilde{H}_2, \tilde{T}_1$, *and* $\tilde{T}_2$, *where a Hopf bifurcation and a Neimark–Sacker bifurcation occur, and these are tangential to* H_1, H_2, T_1, *and* T_2, *respectively.*

The bifurcation analysis suggests the following [50]. Suppose that we vary (δ_1, δ_2) varies along the diagonal line $\{\delta_1 = \delta_2\}$. First, the parameter crosses $\tilde{H}_1$, from which the horizontal oscillation bifurcates. Second, it crosses $\tilde{H}_2$, from which the vertical oscillation bifurcates. Finally, it crosses $\tilde{T}_1$, from which an invariant torus bifurcates. The dynamics on the torus change depending on the parameter value and the phase locking phenomena occur [43, Sect. 7.3]. As shown in Fig. 4.11, there exist parameter regions on the (α_1, α_2)-plane in which the invariant torus supports only a finite number of hyperbolic cycles with the same rotation number. These regions are called Arnold tongues. At the boundary of the Arnold tongues, the limit cycles undergo fold bifurcations.

In conclusion, we can quote [50]: "the transition between periodic and nonperiodic orbits is caused by crossing Arnold tongues, which result from the Hopf–Hopf bifurcation of the trivial equilibrium."

4.5 Some Remarks

The particle model with a single wall was first developed in a master's thesis by Matsumoto [47]. It is observed by numerical simulation that the angles of incidence and reflection are different. Moreover, phase space analysis has been carried out to investigate the occurrence of reflection. In [47] it was shown that the particle on the right-half-plane with an initial velocity $v(0) < 0$ changes the sign of v just once and tends to an equilibrium. The concept of working in (z, v, w)-space was introduced by Matsumoto. The definition of θ_{inc} as the α-limit set was introduced in [15], We have made some estimates of what would be required to ensure that the particle enters each region and converges to an equilibrium as $t \to \pm\infty$. In addition, the present authors have proved $\theta_{\mathrm{inc}} \le \theta_{\mathrm{ref}}$ for $m_0 = 0$ in [15]. Recently, a conjecture on the functional relation for (1.12) has been proposed [51]: as $\delta \to 0$,

$$\theta_{\mathrm{ref}} \approx \theta_{\mathrm{inc}} + \frac{4}{3}\left(1 + \frac{m_0\alpha^2}{m_1 m_2}\right)\frac{\sqrt{\delta m_1}}{\alpha}\sin\theta_{\mathrm{inc}}\cos^2\theta_{\mathrm{inc}}.$$

The functional relation $\theta_{\mathrm{ref}} = F(\theta_{\mathrm{inc}})$ was numerically approximated by Matsumoto [48]. In his master's thesis, asymptotic behavior of particle models on square and rectangular domains was studied by numerical simulation and discrete-time models were derived. We will consider the discrete-time model in the next chapter. The particle model on a square domain was introduced in [16, 47] ([47] refers to a preprint of [16]). In [48] it is found that the particle model with $\delta = 0.06$ and $L = 10$ tends to a square-shaped cycle on a square domain, a rectangular-shaped cycle for $c = 1.2$, and an eight-shaped cycle for $c = 3.0$. In addition, intermittent change between periodic and nonperiodic attractors was found by drawing orbit diagrams for $1 < c < 5$. Morihara [52] also studied the particle model on a rectangular domain by numerical simulation and found many types of cycles. He computed a one-dimensional map that maps the ith reflection point on $x = -L$ to the $(i+1)$th reflection point on the same boundary to estimate the Lyapunov exponent of orbits for $L = 10$ and $c = 1.3$, leading to the conjecture that it is not chaotic but quasiperiodic.

In this chapter, reflection from a single wall and intermittent change between periodic and quasi-periodic attractors were discussed. Section 4.1 was based upon [15], and Sects. 4.2 and 4.4 upon [50]. See [5, 24], and [43] for introductions to dynamical systems and bifurcation theory. See also [22, 23] for details of group-theoretic bifurcation theory, which deals with bifurcations in systems with symmetry.

The orbit diagram of Fig. 4.7 shows additional phenomena of interest. The attractor for $c = 1.85$ and $c = 2.24$ with $L = 10$ seems to be chaotic in the sense that it has a positive Lyapunov exponent [49]. Figure 4.14 shows enlarged pictures of an orbit diagram for $c \in [1.8, 1.81]$ and $c \in [2.224, 2.23]$. We can observe a period-doubling cascade to chaos. Figure 4.15 shows projection onto the (x, w)-plane on the Poincaré section $\{y = 0, \dot{y} > 0\}$. The cases other than $c = 1.3$ seem to be chaotic in the sense of having a positive Lyapunov exponent. As shown in Fig. 4.10, two cycles with different rotation numbers coexist at $c = 1.85$. This means that an invariant torus no longer exists at this parameter value. What causes the invariant torus to collapse in the particle model?

Finally, our way of modeling a particle model on a rectangle may be generalized to polygonal domains. Consider a polygonal region Ω with N vertices $\boldsymbol{c}_0, \ldots, \boldsymbol{c}_{N-1}$ and N edges $\boldsymbol{c}_{j+1} - \boldsymbol{c}_j (j = 0, \ldots, N-1)$, where $\boldsymbol{c}_N = \boldsymbol{c}_0$. Let $\boldsymbol{u} = (x, y) \in \Omega$ and $\boldsymbol{v} = (v, w)$ be the position and velocity of a particle, respectively. Let $\boldsymbol{a}_j$ be the outward normal vector of the edge $\boldsymbol{c}_{j+1} - \boldsymbol{c}_j$. The mirror image of $\boldsymbol{u}$ with respect to the edge $\boldsymbol{c}_{j+1} - \boldsymbol{c}_j$ is given by

$$\boldsymbol{r}_j = \boldsymbol{u} - 2(\boldsymbol{u} - \boldsymbol{c}_j, \boldsymbol{a}_j)\boldsymbol{a}_j,$$

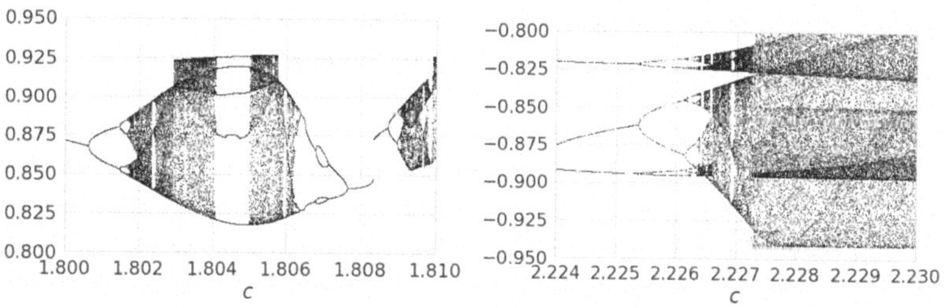

(a) $1.8 < c < 1.81$ (b) $2.224 < c < 2.23$

Fig. 4.14 Enlarged details of Fig. 4.7. Figure adapted from the author's previous work [49]

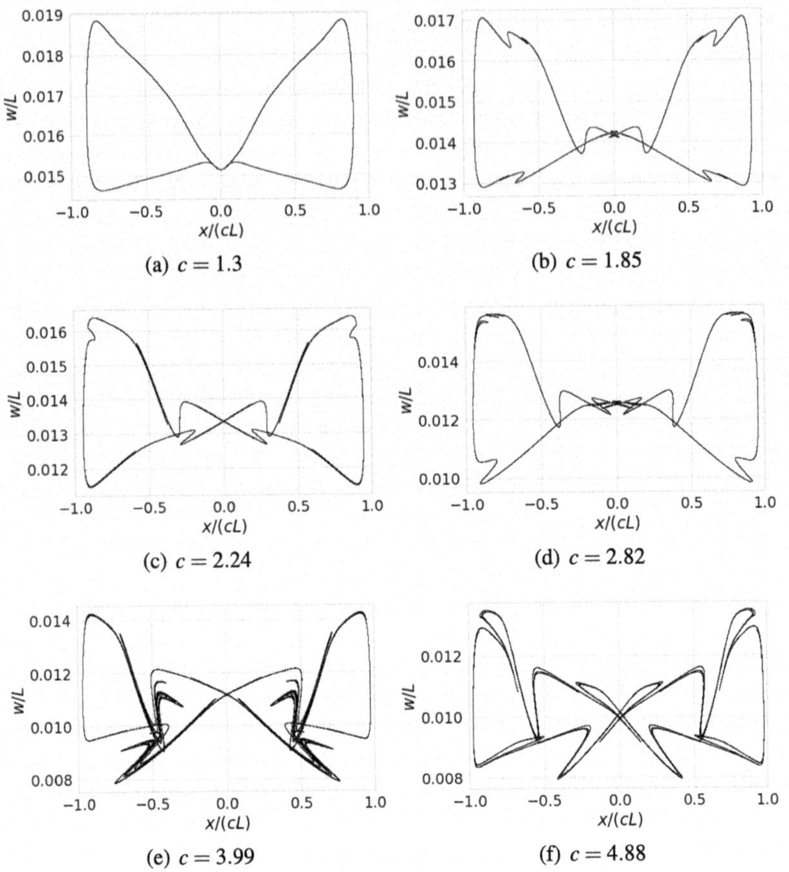

Fig. 4.15 Projection onto (x, w)-plane of the Poincaré section $\{y = 0, \dot{y} > 0\}$

for $j = 0, 1, \ldots, N-1$. Taking interactions with this mirror image into account, we obtain

$$\begin{cases} \dot{\boldsymbol{u}} = \boldsymbol{v} - m_0 \displaystyle\sum_{j=0}^{N-1} d_j^{-3/2} e^{-ad_j} (\boldsymbol{r}_j - \boldsymbol{u}) \\ \dot{\boldsymbol{v}} = \left(\delta - m_1 |\boldsymbol{v}|^2\right) \boldsymbol{v}_j - m_2 \displaystyle\sum_{j=0}^{N-1} d_j^{-3/2} e^{-ad_j} (\boldsymbol{r}_j - \boldsymbol{u}), \end{cases} \tag{4.17}$$

where $d_j = \|\boldsymbol{r}_j - \boldsymbol{u}\|$ for $j = 0, 1, \ldots, N-1$.

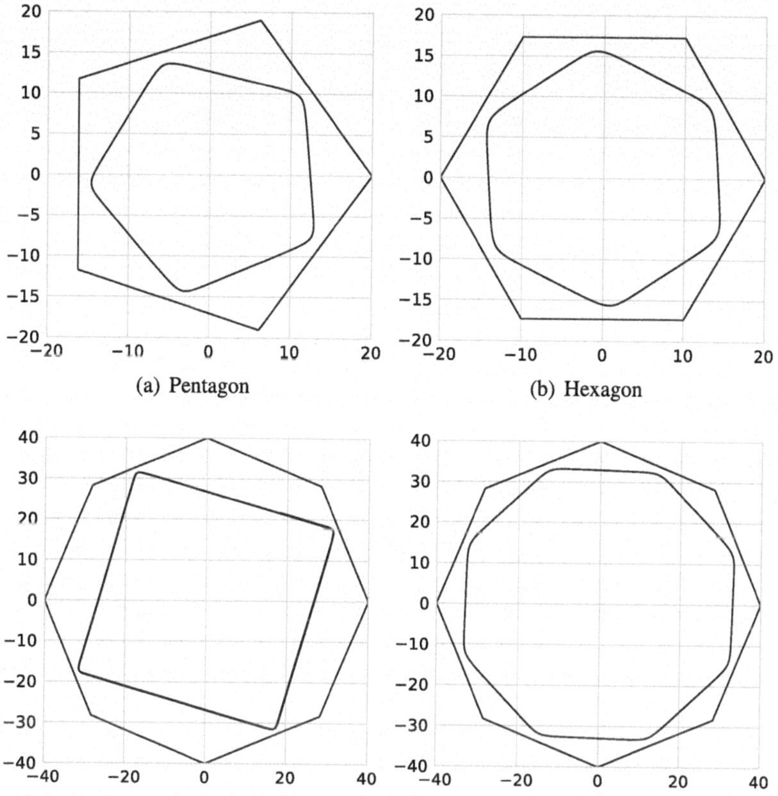

(a) Pentagon (b) Hexagon

(c) Octagon (d) Octagon

Fig. 4.16 Cycles in regular polygons with $\delta = 0.06$ and $m_0 = m_1 = m_2 = a = 1$. $L = 20$ for (**a**, **b**) and $L = 40$ for (**c**, **d**)

If Ω is a regular N-gon, then the system commutes with the action of the dihedral group $\mathbf{D}_N$. Hence, the theory of equivariant Hopf bifurcation with $\mathbf{D}_N$-symmetry is applicable. Figure 4.16 shows limit cycles for some regular N-gon domains. In an octagonal region, square-shaped and octagon-shaped cycles coexist at the same parameter value. Does the limit $N \to \infty$ provide us with a particle model on a circular disk? Figure 4.17 shows two limit cycles on a regular triangle at the same parameter value. Thus, complicated dynamics are expected.

Is this derivation of a particle model valid for a domain Ω with an obtuse angle? See Fig. 4.18, where a particle in Ω repulsively interacts with virtual particles V_1 and V_2. Note that the interaction from V_1 is regarded as the interaction with the edge e_1. Nevertheless, the edge does not lie between the particle and V_1. Does this really represent the interaction with the boundary? Although this observation may be irrelevant if the domain size is sufficiently large, as the interaction terms decay exponentially, we need to consider the modeling more carefully. A mathematical

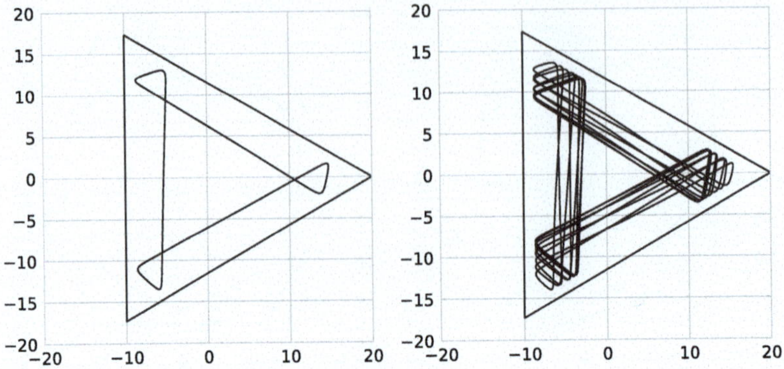

Fig. 4.17 Cycles on a regular triangle. Parameters are $\delta = 0.05$, $L = 20$, $m_0 = m_1 = m_2 = a = 1$

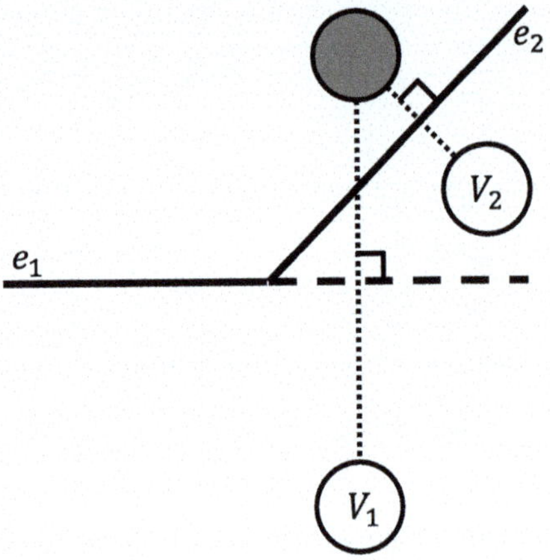

Fig. 4.18 Reflection near an obtuse angle

scheme for modeling a particle model in a bounded region including rectangles and polygons is currently under development. Koyano et al. [41] has studied camphor motion in a disk from the viewpoint of theoretical physics. On the basis of a PDE model, a particle model can be derived from a solution of an elliptic equation. This has not, however, been achieved for the general case as yet.

Chapter 5
Discrete-Time Models

Abstract A discrete-time model is introduced to discuss the existence and stability of periodic orbits. In particular, square-shaped and parallelogram-shaped cycles on square and more generically rectangular domains are analyzed. In addition, a figure-eight-shaped cycle on a rectangular domain is also studied. A discrete-time model is formally derived in the limit where the domain size tends to infinity. Numerical simulation suggests that the discrete-time model approximates the corresponding particle model when the domain size is sufficiently large. An observable gap between the orbits for these two models in cases where the domain is of medium scale is also discussed.

5.1 Introduction of a Discrete-Time Model

We have discussed billiard-like motions on a rectangular domain for the moving boundary model and the particle model. We would like to obtain a deeper understanding of the limit cycles. For this purpose, we introduce a discrete-time model. First of all, let us observe the particle model on a square domain $[-L, L] \times [-L, L]$ by numerical simulation.

How do orbits depend on the domain size L? Recall Figs. 1.21 and 1.22, which show square-shaped trajectories and the time dependence of speed for $L = 10, 10^2$, and 10^3, where each figure is scaled by L. For larger domain sizes L, reflection points are relatively closer to the boundary. While the speed diminishes temporarily when the reflection occurs, it recovers after the reflection. It should be noted that the elapsed time during reflection becomes relatively shorter as L increases.

Suppose that the billiard table is huge relative to the camphor disk size. In the limit of $L \to \infty$, we may assume that the particle collides with the boundary and reflects instantaneously because of the observations above. Although instant reflection is a property shared with conventional forms of mathematical billiards, the reflection angle remains different from that of incidence in our case. As we have discussed in

T. Miyaji et al., *A Billiard Problem in Nonlinear Dissipative Systems*,
Surveys and Tutorials in the Applied Mathematical Sciences 18,
https://doi.org/10.1007/978-981-95-7561-9_5

Chap. 4, the reflection angle is determined by a prescribed function $\theta_{\mathrm{ref}} = F(\theta_{\mathrm{inc}})$. Thus we arrive at a discrete-time dynamics model.

Here, we summarize numerical methods of computing an orbit on a rectangle with aspect ratio c. See Fig. 5.1. Let $v_0, v_1, v_2, v_3 \in \mathbb{C}$ be the vertices of the rectangular domain D with aspect ratio c, say,

$$v_0 = 1, \quad v_1 = \exp(\mathrm{i}\phi), \quad v_2 = -1, \quad v_3 = \exp(\mathrm{i}(\pi + \phi)),$$

where $\phi = 2\arctan(1/c)$. Edges of the rectangular domain are:

$$e_0 = v_1 - v_0, \quad e_1 = v_2 - v_1, \quad e_2 = v_3 - v_2, \quad e_3 = v_0 - v_3.$$

Let n_0, n_1, n_2, n_3 be the outward normal vectors of the domain given by

$$n_j = \exp(-\mathrm{i}\pi/2)e_j, \quad (j = 0, 1, 2, 3).$$

Suppose that we have a reflection point $z_0 \in \partial D$ and a direction vector $w_0 \in \mathbb{C}$, $|w_0| = 1$. The next reflection point z_1 and the next direction vector w_1 are determined in the following manner. Note that z_1 is on the half straight line $\{z_0 + tw_0 \mid t > 0\}$. Let $\tilde{z} = z_0 + sw_1$ be the point on the unit circle with $s > 0$. There exists j such that

$$\arg v_j \leq \arg \tilde{z} < \arg v_{j+1}.$$

The point z_1 lies on the edge e_j. Solving the linear equation $z_0 + tw_1 = v_j + ue_j$ for (t, u), we obtain $z_1 = z_0 + tw_0$ (Fig. 5.1b). The angle of incidence is given by

$$\theta_{\mathrm{inc}} = \arccos \frac{\mathrm{Re}(n_j \overline{w_0})}{|w_0||n_j|},$$

and that of reflection is $\theta_{\mathrm{ref}} = F(\theta_{\mathrm{inc}})$. Let $\theta = \arccos(\mathrm{Re}(e_j\overline{w_0})/(|w_0||e_j|))$ be the angle between w_0 and e_j. The velocity vector after reflection is given by

$$w_1 = -w_0 \exp\left(\mathrm{i}\,(\theta_{\mathrm{inc}} + \theta_{\mathrm{ref}})\,\mathrm{sgn}\left(\theta - \frac{\pi}{2}\right)\right).$$

Figures are drawn by applying a similarity transformation so that the vertices of the rectangle are located at $(c, 1)$, $(-c, 1)$, $(-c, -1)$, and $(c, -1)$.

Figure 5.2 shows trajectories of the discrete-time model on square, where the reflection rule was obtained from the particle model with a single wall (4.2) with $m_0 = m_1 = m_2 = a = 1$ and $\delta = 0.06$ (see Fig. 5.3). We can observe vertical and horizontal oscillations and a square-shaped cycle. Figure 5.2c shows the horizontal oscillation, and Fig. 5.2a, b, d are trajectories that converge to the square-shaped cycle. In contrast to the particle model, there exist families of vertical and horizontal oscillations. Here, vertical oscillation refers to a periodic orbit perpendicular to the long side of a rectangle, and horizontal oscillation refers to one perpendicular to the

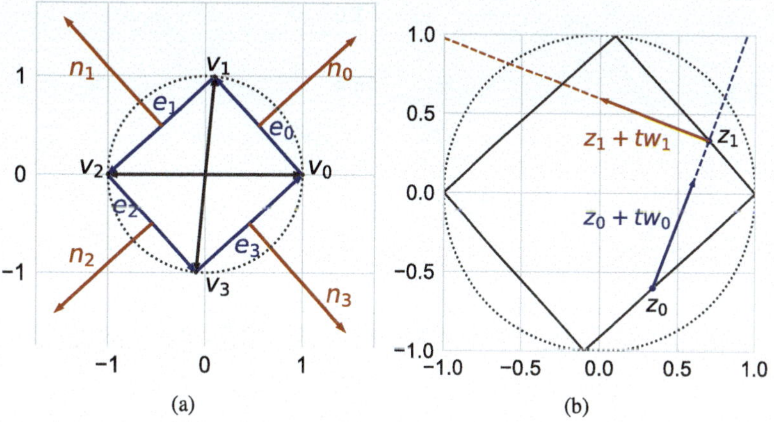

Fig. 5.1 Illustration of our numerical method for the discrete model. **a** Interpretation of the complex numbers as geometric vectors. **b** Reflection points

short side. Indeed, if the incident angle is zero, then the trajectory is a line segment between two edges, regardless of reflection points. However, these oscillations are unstable. Even if the initial velocity vector is almost perpendicular to the boundary of the table, it eventually converges to a square-shaped cycle as shown in Fig. 5.2d.

Figure 5.4 shows attractors in a rectangular domain for some aspect ratio c. Compare these figures with Fig. 4.8. Both models have qualitatively similar attractors for $c = 1.38, 1.46, 1.6, 1.9$. On the other hand, there are noticeable differences for $c = 1.705$ and 1.75. Since the discrete model assumes $L \to \infty$, it does not necessary represent trajectories of particle model for small L.

To investigate the discrete model in detail, we impose some assumptions on the angle relation F. As we have seen in the previous chapter, the function F exists for the particle model with a single wall. We have shown that $F(\theta) > \theta$ for a special case where $m_0 = 0$. Although numerical simulation suggests that it holds even for $m_0 > 0$, it is difficult to obtain a formula of F in a closed form. We may assume that F satisfies the following properties:

Assumption 5.1 *1. F is continuous,*
2. $F(0) = 0, \quad F(\frac{\pi}{2}) = \frac{\pi}{2}$,
3. $F(\theta) > \theta$ for any $\theta \in (0, \pi/2)$,
4. F is strictly monotone increasing.

If F is a continuously differentiable function with Assumption 5.1, then it must satisfy

$$F'(0) \geq 1, \quad F'\left(\frac{\pi}{2}\right) \leq 1.$$

In fact, numerical simulations suggest that the angle relation for the particle model satisfies the following properties:

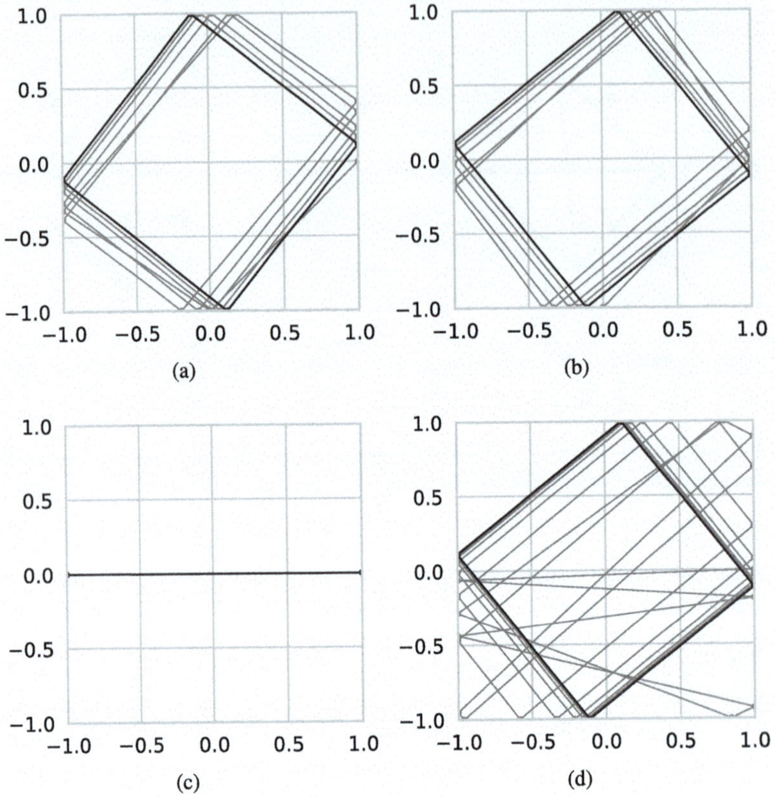

Fig. 5.2 Trajectories of the discrete-time model on square

Fig. 5.3 Angle relation obtained from the particle model with $\delta = 0.06$, $a = m_0 = m_1 = m_2 = 1$. Figure adapted from the author's previous work [50, Fig. 6]

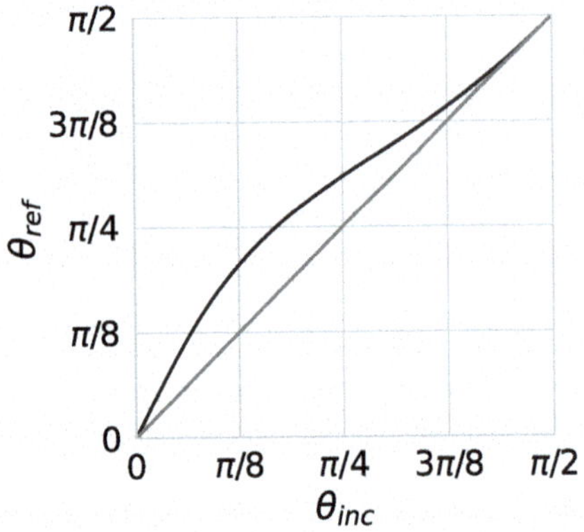

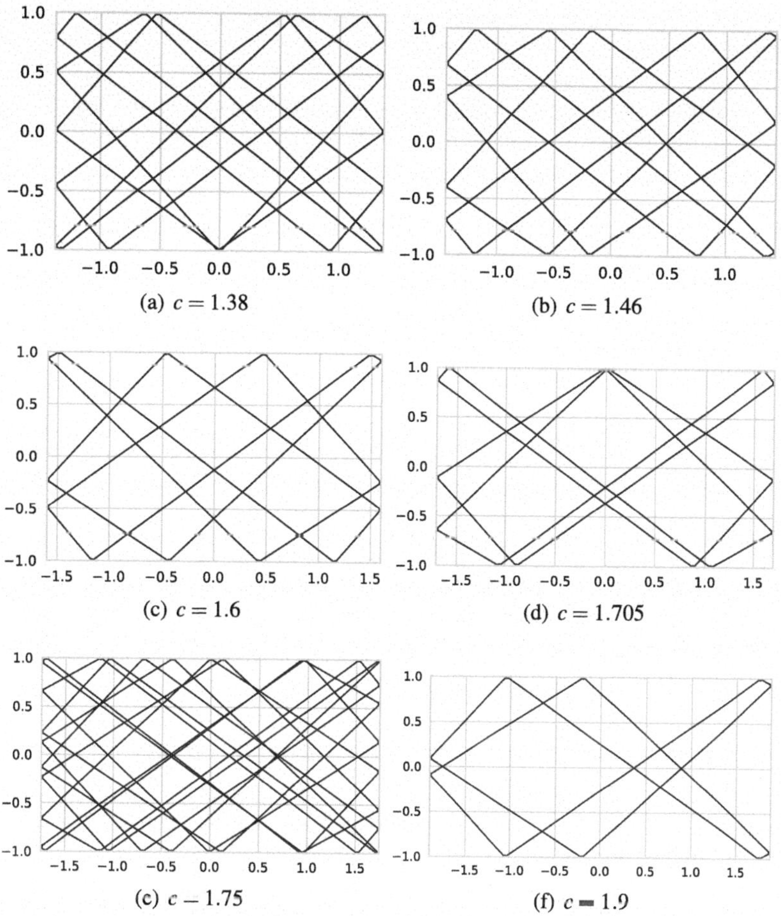

(a) $c = 1.38$ (b) $c = 1.46$

(c) $c = 1.6$ (d) $c = 1.705$

(e) $c = 1.75$ (f) $c = 1.9$

Fig. 5.4 Cycles of the discrete model on a rectangle. The angle relation is shown in Fig. 5.3. Figure adapted from the author's previous work [49]

Assumption 5.2 *F is continuously differentiable and satisfies*

$$F'\left(\frac{\pi}{2}\right) = 1.$$

Assumption 5.3 *F is continuously differentiable and satisfies*

$$F'(0) > 1.$$

Although our starting point is camphor disk motion, it will be useful for us to study dynamics for more general reflection rules. We study dynamical systems assuming that F satisfies the conditions above.

The following theorem is evident.

Theorem 5.1 *If F satisfies Assumption 5.1, then families of vertical and horizontal oscillations exist and they are unstable.*

The dynamics of the angle of reflection is governed by iteration of F when the trajectory is sufficiently close to such oscillations. Hence, the reflection angle diverges from an initial angle close to zero unless the initial velocity vector is exactly perpendicular to the edge.

We study the existence and stability of a square-shaped cycle in the next section.

5.2 Square Domain

We consider the existence and stability of a square-shaped cycle in a square domain. Portions of this section are adapted from [15], used with permission of the American Institute of Mathematical Sciences, from Ei et al. [15]; permission conveyed through Copyright Clearance Center, Inc.

Suppose that the particle hits all the edges of a square $ABCD$ in turn as shown in Fig. 5.5. Starting from the edge AB, it visits edges BC, CD and DA, and returns to AB. A reflection point divides an edge into an $x : (1 - x)$ ratio. We regard x as the position of reflection. If the angle of incidence is θ_n and the position is x_n in the nth reflection, then x_{n+1} and θ_{n+1} are given by

$$x_{n+1} = (1 - x_n) \cot F(\theta_n), \qquad \theta_{n+1} = \frac{\pi}{2} - F(\theta_n).$$

Therefore, we consider a two-dimensional dynamical system defined by $\mathscr{S}(x, \theta) = (X(x, \theta), G(\theta))$, where

$$X(x, \theta) = (1 - x) \cot F(\theta), \qquad G(\theta) = \frac{\pi}{2} - F(\theta). \tag{5.1}$$

This dynamical system can describe the motion of the particle in the square domain as long as $0 < X(x, \theta) < 1$ is satisfied. Note that our model does not distinguish between clockwise and anticlockwise motion.

First, we show the existence of a square-shaped cycle.

Theorem 5.2 *([15]) If F satisfies Assumption 5.1, then $\mathscr{S}$ has the unique fixed point (x^*, θ^*) in $(0, 1) \times (0, \pi/2)$ defined by*

$$x^* = \frac{1}{1 + \tan F(\theta^*)} = \frac{\tan \theta^*}{1 + \tan \theta^*}, \tag{5.2}$$

where θ^* is the fixed point of G. Accordingly, there exists a unique square-shaped periodic orbit.

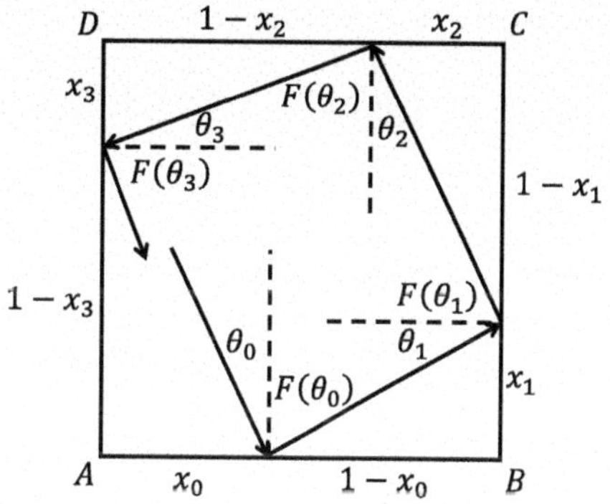

Fig. 5.5 Illustration of discrete model. Figure adapted from [15, Fig. 8], used with permission of the American Institute of Mathematical Sciences, from Ei et al. [15]; permission conveyed through Copyright Clearance Center, Inc.

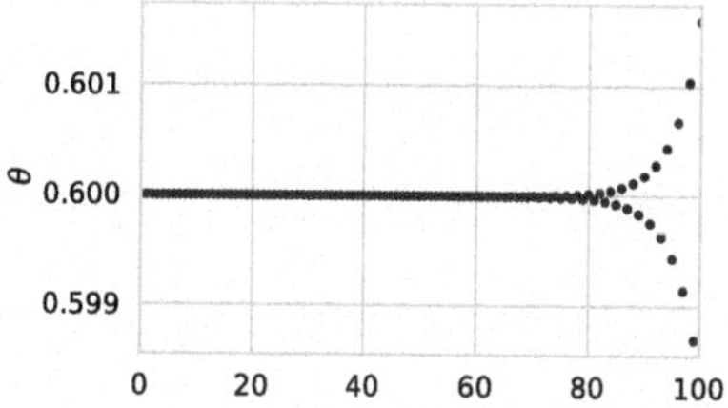

Fig. 5.6 A time sequence of iteration of G for quadratic function in Example 5.3

Proof Since $G(0) - 0 = \pi/2$ and $G(\pi/2) - \pi/2 = -\pi/2$ hold, the intermediate value theorem implies that G has a fixed point. In addition, the monotonicity of F implies uniqueness. Let θ^* be the fixed point of G. Since we have $\theta^* < F(\theta*)$ and

$$\tan\theta^* = \tan G(\theta^*) = \tan\left(\frac{\pi}{2} - F(\theta^*)\right) = \cot F(\theta^*),$$

we obtain

$$0 < \theta^* < \frac{\pi}{4} < F(\theta^*) < \frac{\pi}{2} \tag{5.3}$$

and $x^* \in (0, 1)$. Moreover, $\theta^* - G(\theta^*)$ is equivalent to $\theta^* + F(\theta^*) - \pi/2$, which means that the cycle forms a square. □

Example 5.3 Let us consider the angular relation given by

$$F(\theta) = \frac{2}{\pi}\theta(\pi - \theta).$$

This satisfies Assumption 5.1. The fixed point of $G(\theta) = \pi/2 - F(\theta)$ is $\theta^* = (3 - \sqrt{5})\pi/4$. Figure 5.6 shows a time-series of angular dynamics with the initial data $\theta = \theta^+ + 10^{-12}$. The square-shaped cycle exists but it seems unstable.

Example 5.3 is important! The property $F(\theta) > \theta$ alone does not imply the stability of a square-shaped cycle. The next theorem establishes a criterion for the linear stability of a square-shaped cycle.

Theorem 5.4 *([15]) Suppose that F satisfies Assumptions 5.1 and 5.2. Let (x^*, θ^*) be the fixed point of $\mathscr{S}$. The linearized eigenvalues of $\mathscr{S}$ at (x^*, θ^*) are $-\cot F(\theta^*)$ and $-F'(\theta^*)$. If $F'(\theta^*) < 1$, then the square-shaped cycle is stable. If $F'(\theta^*) > 1$, then it is unstable.*

Proof The linearized matrix of $\mathscr{S}$ about the fixed point (x^*, θ^*) is an upper triangular matrix given by

$$\begin{pmatrix} -\cot F(\theta^*) & -(1-x^*)F'(\theta^*)\csc^2 F(\theta^*) \\ 0 & -F'(\theta^*) \end{pmatrix}.$$

Its eigenvalues are $-\cot F(\theta^*)$ and $-F'(\theta^*)$. By (5.3), we obtain $|-\cot F(\theta^*)| < 1$. Hence, the stability of the fixed point (x^*, θ^*) is determined by the modulus of $-F'(\theta^*)$. □

Example 5.5 *(Angle relation of a particle model)* Let us apply the theorems above to the angle relation shown in Fig. 5.3. We interpolated 129 equally distributed sample points by means of cubic spline, and numerically computed the fixed point of G and its derivative. We obtained $\theta^* \approx 0.673$, $x^* \approx 0.724$, and $F'(\theta^*) \approx 0.393$. We have $x^* \in (0, 1)$ and $|F'(\theta^*)| < 1$. Therefore the square-shaped cycle exists, and it is stable.

We study the properties of G. Let $I_1 = [0, \theta^*]$ and $I_2 = [\theta^*, \pi/2]$. Since G is monotone decreasing, we have $G(I_1) = I_2$ and $G(I_2) = I_1$. Therefore, G cannot have periodic points with odd minimal period. Moreover, the following proposition holds.

Proposition 5.1 *1. G^2 is strictly monotone increasing in $\theta \in (0, \pi/2)$.*

2. G cannot have periodic points whose minimal period is greater than or equal to four.

3. G^2 has at least three fixed points

$$0, \quad \frac{\pi}{2}, \quad \theta^*,$$

where θ^* is the fixed point of G. Moreover, if $F'(\pi/2) = 1$ and $F'(0) > 1$, then the periodic orbit $\{0, \pi/2\}$ is unstable for G.

Proof By elementary calculus and Assumption 5.1, we have

$$(G^2)'(\theta) = G'(G(\theta))G'(\theta) = F'(\pi/2 - F(\theta))F'(\theta) > 0.$$

Therefore, any sequence generated by iteration of G^2 is monotone, and G^2 cannot have periodic points other than fixed points. In particular, G does not have periodic points of minimal period 4. This proposition follows from Sharkovskii's theorem.

Theorem 5.2 shows the existence of θ^*. Moreover, we obtain

$$G(0) = \frac{\pi}{2} - F(0) = \frac{\pi}{2}, \quad G\left(\frac{\pi}{2}\right) = \frac{\pi}{2} - F\left(\frac{\pi}{2}\right) = 0.$$

The linear stability of these periodic points is determined by the derivative of G^2. It is evident that

$$(G^2)'(0) = (G^2)'(\pi/2) = F'(\pi/2)F'(0).$$

If we assume $F'(\pi/2) = 1$ and $F'(0) > 1$, then we obtain

$$(G^2)'(0) = (G^2)'(\pi/2) = F'(0) > 1.$$

□

Theorem 5.6 *Suppose that F satisfies Assumption 5.1 and that G has a nontrivial periodic orbit $\{\theta_0, \theta_1\}$, where $\theta_1 = G(\theta_0) \neq \theta_0$. Let x_0 and x_1 be given by*

$$x_0 = \frac{\tan\theta_0(1 - \tan\theta_1)}{1 - \tan\theta_0 \tan\theta_1}, \quad x_1 = \frac{\tan\theta_1(1 - \tan\theta_0)}{1 - \tan\theta_0 \tan\theta_1}. \tag{5.4}$$

If $0 < x_0 < 1$ and $0 < x_1 < 1$ hold, then there exists a parallelogram-shaped cycle.

Therefore, if there is a convex quadrilateral cycle, then it must be a square or a parallelogram in the more general instance.

Proposition 5.2 *Under Assumption 5.1, the following statements are true.*

1. *If G has a globally asymptotically stable fixed point θ^*, then G^2 has only trivial fixed points and $F(0)F(\pi/2) \geq 1$ holds.*
2. *If G has only trivial fixed points and $F'(\theta^*) < 1$, then θ^* is globally asymptotically stable for G.*

Example 5.7 (*Angle relation of a particle model*) Let us consider the angle relation shown in Fig. 5.3 again. Figure 5.7 plots G and G^2. It turns out that the fixed point of G is globally asymptotically stable for G.

Example 5.8 (*Quadratic function* [15]) We return to the quadratic case. If F is a quadratic function satisfying Assumption 5.1, then it is given by

$$F(\theta) = \theta\,(1 + \pi a - 2a\theta),$$

where $0 < a \leq 1/\pi$. In this case, Assumption 5.2 is not satisfied. Since we have

$$F'(\theta) = 1 + \pi a - 4a\theta > 1$$

for any $a \in (0, 1/\pi]$ and $\theta \in (0, \pi/4)$, the theorems above imply that the square-shaped cycle is unstable. In addition, no parallelogram cycle exists. Indeed, as

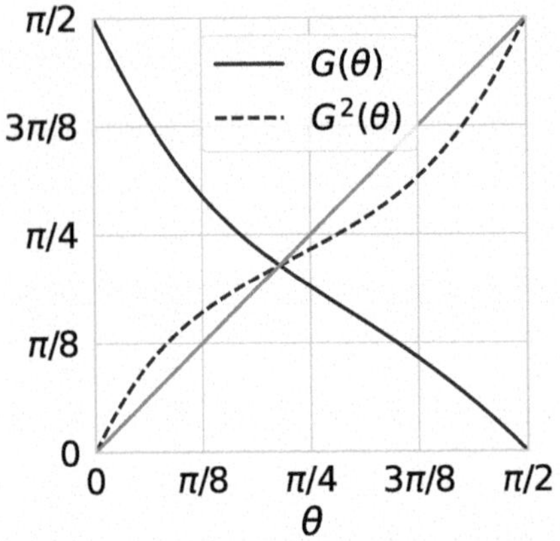

Fig. 5.7 Graphs of G and G^2. Particle model with $m_0 = m_2 = 1$, $\delta = 0.06$. θ^* is globally asymptotically stable for G^2

$$G^2(\theta) - \theta = -a^2\theta(\pi - 2\theta)\left[4a\theta^2 - 2(a\pi + 2)\theta + \pi\right]$$

holds, there is no nontrivial periodic point of period 2.

Since we have

$$(G^2)'(0) = (G^2)'(\pi/2) = F'(\pi/2)F'(0) = 1 - a^2\pi^2 < 1,$$

0 and $\pi/2$ are stable for G^2 (see Fig. 5.11). The orbit, however, does not necessarily satisfy the assumption of $\mathscr{S}$. That is, after reflection at edge AB of a square $ABCD$, the next reflection may occur at edge CD rather than edge BC. In such a case, the orbit is not determined by the angular dynamics alone.

From the theorem above, the instability of the square-shaped cycle occurs when $G(\theta^*) = -F'(\theta^*)$ passes through -1 as a parameter changes. This suggests a possibility of a period-doubling bifurcation. We apply Theorem 2.25 for studying a period doubling-bifurcation ([24, Theorem 3.5.1]).

Example 5.9 (*Cubic function*) Suppose that $F(\theta)$ is a cubic function satisfying Assumption 5.1. F is given by

$$F(\theta) = \theta + \theta(\pi - 2\theta)(a\theta - b), \tag{5.5}$$

where a and b are real parameters. By Assumption 5.1(3.), a and b satisfy $b/a > \pi/2$ or $b/a < 0$. In addition, since F is monotonically increasing, a and b are contained in $D_1 \cup D_2 \cup D_3$, where

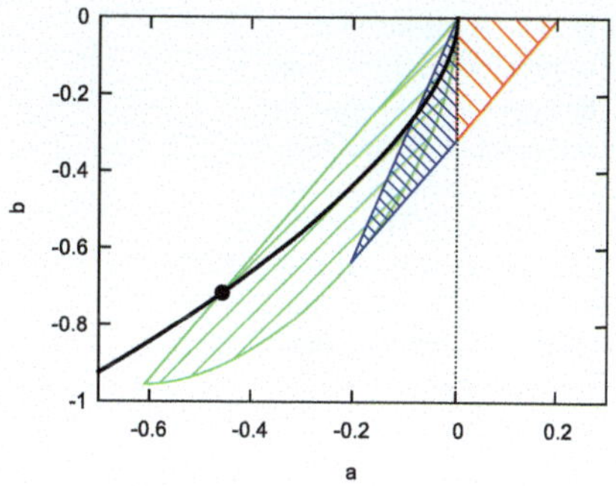

Fig. 5.8 Regions of (a, b) in which (5.5) satisfies Assumption 5.1 and the curve of period-doubling bifurcation given by (5.7). D_1, D_2, and D_3 are indicated by red, green, and blue, respectively

$$D_1 = \left\{(a, b) \mid 0 < a \le \frac{2}{\pi^2}, \frac{a\pi^2 - 2}{2\pi} \le b \le 0\right\}$$

$$D_2 = \left\{(a, b) \mid -\frac{2}{\pi^2} \le a < 0, \frac{a\pi^2 - 2}{2\pi} \le b \le a\pi\right\}$$

$$D_3 = \left\{(a, b) \mid -\frac{6}{\pi^2} \le a < 0, \frac{a\pi - \sqrt{-3a^2\pi^2 - 24a}}{4} \le b \le \frac{a\pi}{2}\right\}.$$

The square-shaped cycle has -1 as an eigenvalue if the following equations hold:

$$\begin{cases} \frac{\pi}{2} - F(\theta) = \theta, \\ -F'(\theta) = -1. \end{cases} \tag{5.6}$$

Solving this linear equation for (a, b), we obtain

$$a(\theta) = -\frac{(\pi - 4\theta)^2}{2\theta^2(\pi - 2\theta)^2}, \quad b(\theta) = -\frac{(\pi - 4\theta)(\pi - 3\theta)}{\theta(\pi - 2\theta)^2}, \tag{5.7}$$

where (a, b) is parametrized by θ for $0 < \theta < \pi/4$ because the fixed point of the angular dynamics must lie in $(0, \pi/4)$. The fixed point of G undergoes a period-doubling bifurcation when the parameter crosses the curve $(a(\theta), b(\theta))$. Figure 5.8 shows the curve and the regions D_1, D_2 and D_3.

Example 5.10 (*Cubic function 2*) Next, we add Assumption 5.2, that is, $F'(\pi/2) = 1$. In this case, we have

$$F(\theta) = \theta + a\theta(\pi - 2\theta)^2, \tag{5.8}$$

where a is contained in the open interval $(0, 3/\pi^2)$. Note that a is different from that in (5.5). This is the simplest case satisfying Assumptions 5.1 and 5.2. In addition, it also satisfies Assumption 5.3. The stability of the fixed point of G depends on the parameter a. Since F is cubic, it is possible to compute the fixed point. However, it

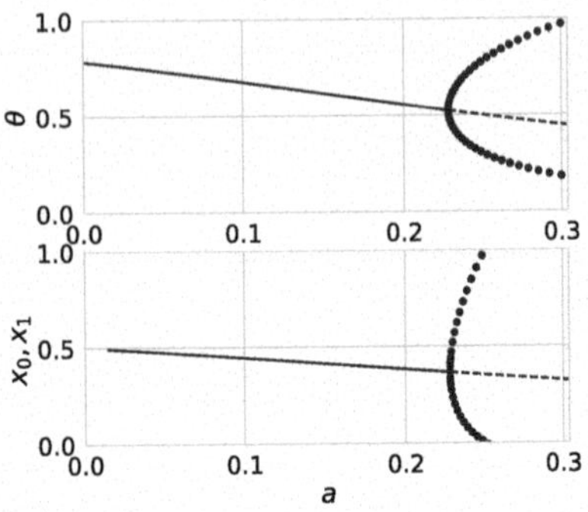

Fig. 5.9 A bifurcation diagram for the cubic case. The upper figure shows the fixed point of G, and the lower figure shows x_0 and x_1 given by (5.4). The solid real line, dashed line, and dots indicate stable fixed point, unstable fixed point, and stable periodic point of G, respectively

is not easy to understand its properties from the formula. We compute the value of a at which the fixed point loses stability. Solving the system

$$\begin{cases} G(\theta) = \theta, \\ F'(\theta) = 1, \end{cases}$$

we obtain $\theta = \pi/6$ and $a = 9/(4\pi^2)$. The fixed point of G is stable for $a \in (0, 9/(4\pi^2))$, and it undergoes a period-doubling bifurcation at $a = 9/(4\pi^2)$. On the other hand, G^2 has nontrivial fixed points if $a \in (9/(4\pi^2), 3/\pi^2)$. See Figs. 5.12 and 5.13. Figure 5.9 shows a bifurcation diagram. Figure 5.10 shows attractors for $a = 0.2, 0.24$ and 0.29. Although G has a stable periodic orbit for $a \in (9/(4\pi^2), 3/\pi^2)$, a parallelogram-shaped cycle does not necessarily exist. At $a = 2(\sqrt{5} - 1)/\pi^2$, the periodic points of G are $\theta_0 = \pi/4$ and $\theta_1 = (3 - \sqrt{5})\pi/8$. Note that $(x_0, x_1) = (1, 0)$ if $\theta_0 = \pi/4$, where x_0 and x_1 are given by (5.4). Therefore, a parallelogram-shaped cycle exists if and only if $a \in (9/(4\pi^2), 2(\sqrt{5} - 1)/\pi^2)$. Otherwise, a stable eight-shaped cycle appears (Fig. 5.10(c)).

Example 5.11 (*Quartic function* [15]) Suppose that F is a quartic function satisfying Assumptions 5.1–5.3. We denote F by

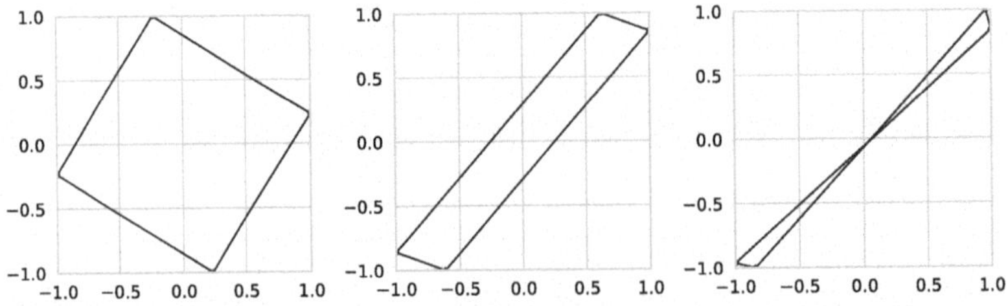

(a) $a = 0.20$ (b) $a = 0.24$ (c) $a = 0.29$

Fig. 5.10 Limit cycles for a cubic function

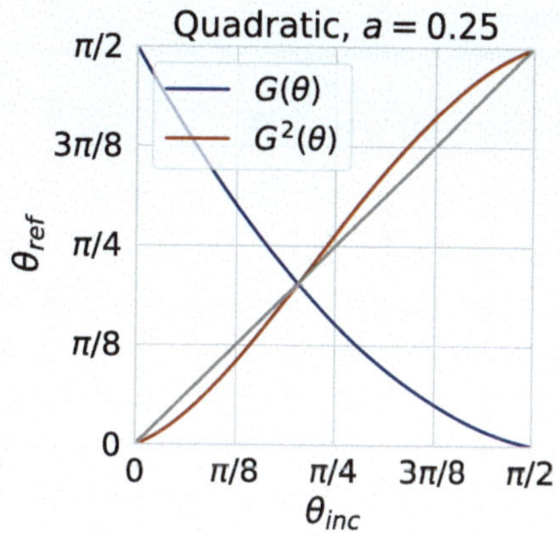

Fig. 5.11 Graphs of G and G^2. Quadratic function with $a = 0.25$. θ^* is unstable for G^2

$$F(\theta) = \theta + \theta(\pi - 2\theta)^2(a\theta + b),$$

where a and b are real numbers. By assumption, it is necessary that a and b satisfy

$$\pi a \geq -2b, \quad b > 0.$$

Although the monotonicity of F imposes another condition, we omit that detail. Theorem 5.2 implies the unique existence of a square-shaped cycle. In a similar manner to the cubic case, its stability depends on the parameter (a, b). The system of equations

$$G(\theta) = \theta, \qquad F'(\theta) = 1$$

is linear with respect to (a, b). By solving it, we obtain a curve $\{(a(\theta), b(\theta)) \mid 0 < \theta < \pi/4\}$,

$$a(\theta) = -\frac{(\pi - 4\theta)(\pi - 6\theta)}{2\theta^2(\pi - 2\theta)^3}, \qquad b(\theta) = \frac{(\pi - 4\theta)^2}{\theta(\pi - 2\theta)^3}. \tag{5.9}$$

If $(a, b) = (a(\theta), b(\theta))$ for $\theta \in (0, \pi/4)$, then the square-shaped cycle has the multiplier -1. One can verify that a supercritical period-doubling bifurcation occurs if the parameter crosses this curve transversally. As a result, a parallelogram-shaped cycle bifurcates.

Example 5.12 Morihara [52] has discovered an interesting example. Consider $F(\theta) = \pi/2 - F_4(\pi/2 - \theta)$ with $F_4(\theta) = 2\theta^2/\pi + (1 - \cos 4\theta)/8$. G has unstable periodic points of minimal period 2, and a stable periodic orbit $\{0, \pi/2\}$. Nevertheless, its typical trajectory converges to a square-shaped cycle. In this example, the trajectory is not determined by iteration of $\mathscr{S}$ alone.

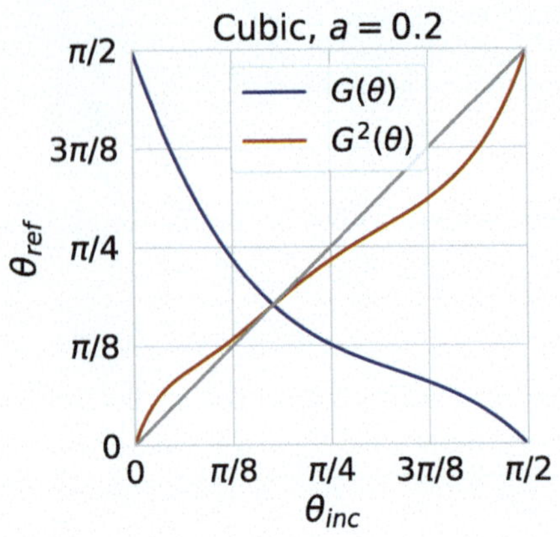

Fig. 5.12 Graphs of G and G^2. Cubic function with $a = 0.2$. θ^* is globally asymptotically stable for G^2

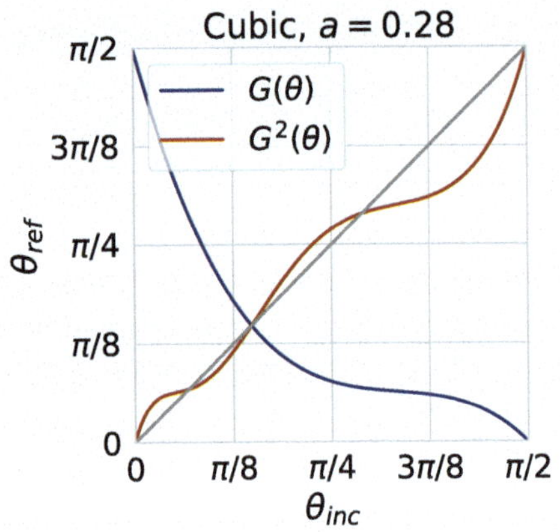

Fig. 5.13 Graphs of G and G^2. Cubic function with $a = 0.28$. θ^* is unstable but nontrivial fixed points are stable for G^2

5.3 Rectangular Domain

In this section, we consider the case where the domain is a rectangle $ABCD$ with the aspect ratio c, where $AB : BC = c : 1$, $AB = CD$, and $BC = AD$. We need only consider $c > 1$ because $c : 1 = 1 : c^{-1}$. We know that families of horizontal and vertical oscillations exist and they are unstable.

First, we study a rectangular cycle. Consider a billiard ball that hits all the edges in turn as shown in Fig. 5.14. Suppose that a reflection site divides one of the long sides into a $x_0 : (c - x_0)$ ratio and that the angle of incidence is θ_0. Now, the angle of incidence at the nth reflection is determined by $\theta_n = G^n(\theta_0)$ in the same way as in the previous section. At the same time, the reflection site at the nth reflection is given by

$$x_n = \begin{cases} (c - x_{n-1}) \cot F(\theta_{n-1}) & \text{if } n \text{ is odd} \\ (1 - x_{n-1}) \cot F(\theta_{n-1}) & \text{if } n \text{ is even.} \end{cases}$$

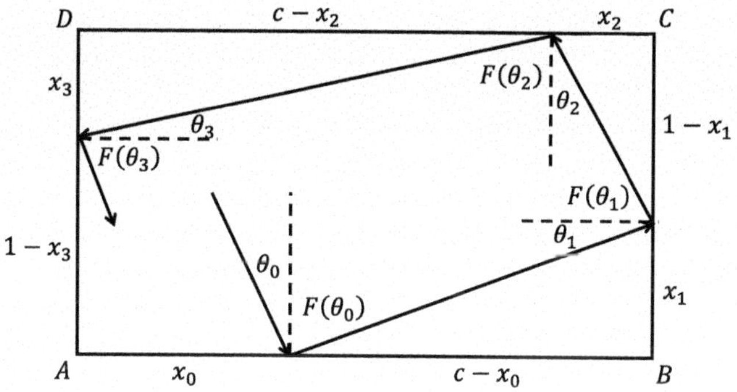

Fig. 5.14 Illustration of the map $\mathscr{R}$

We can define a map $\mathscr{R}(x, \theta) = (Y(x, \theta), G^2(\theta))$ from one of the long sides to another long side by

$$Y(x, \theta) = (1 - (c - x) \cot F(\theta)) \cot F(G(\theta)).$$

If $\mathscr{R}$ has a fixed point (x^*, θ^*) such that $G(\theta^*) = \theta^*$, $0 < x^* < c$, and

$$0 < (c - x^*) \cot F(\theta^*) < 1,$$

then there exists a rectangular cycle.

Theorem 5.13 *Suppose that F satisfies Assumption 5.1. Let θ^* be the fixed point of the angular dynamics G. A rectangular cycle exists in $[0, c] \times [0, 1]$ if and only if*

$$\tan \theta^* < c < \cot \theta^*. \tag{5.10}$$

The cycle is unique if it exists. If F is continuously differentiable and $F'(\theta^*) < 1$, then it is stable, while it is unstable if $F'(\theta^*) > 1$.

Proof Let $x_j (j = 0, 1, 2, 3)$ be positions of reflection given by

$$x_1 = (c - x_0) \tan \theta^*$$
$$x_2 = (1 - x_1) \tan \theta^*$$
$$x_3 = (c - x_2) \tan \theta^*.$$

A rectangular cycle exists if and only if $0 < x_0 = x_2 < c$ and $0 < x_1 = x_3 < 1$. Suitable x_0 and x_1 are given by

$$x_0 = \frac{(1 - c \tan \theta^*) \tan \theta^*}{1 - \tan^2 \theta^*}, \quad x_1 = \frac{(c - \tan \theta^*) \tan \theta^*}{1 - \tan^2 \theta^*}.$$

Here, $0 < \tan\theta^* < 1$ holds because $0 < \theta^* < \pi/4$. Therefore the sufficient and necessary condition is (5.10). Similar to Theorem 5.4, the linearized eigenvalues of (x^*, θ^*) are

$$\frac{\partial Y}{\partial x}(x^*, \theta^*) = \cot^2 F(\theta^*), \quad \frac{dG^2}{d\theta}(\theta^*).$$

We have $\cot^2 F(\theta^*) < 1$ because $\theta^* < \pi/4 < F(\theta^*)$. Therefore, the stability of (x^*, θ^*) is determined by $F'(\theta*)$. □

A cycle can be formed when the fixed point of $\mathscr{R}$ is given by a periodic orbit of G with minimal period 2. In such a case, the cycle forms a parallelogram.

Theorem 5.14 *Let $\{\theta_0, \theta_1\}$ be a periodic orbit with minimal period 2 for G. That is, $\theta_1 = G(\theta_0)$ and $\theta_1 \neq \theta_0$ hold. A parallelogram-shaped cycle exists in $[0, c] \times [0, 1]$ if and only if*

$$\tan\theta_0 < c < \cot\theta_1.$$

The parallelogram-shaped cycle is stable if $\{\theta_0, \theta_1\}$ is stable. Otherwise, it is unstable.

Proof If the fixed point of $\mathscr{R}$ is given by a periodic orbit $\{\theta_0, \theta_1\}$ of G satisfying $\theta_1 = G(\theta_0) \neq \theta_0$, then the resultant cycle forms a parallelogram. The reflection points are given by

$$x_0 = \frac{\tan\theta_0(1 - c\tan\theta_1)}{1 - \tan\theta_0\tan\theta_1}, \quad x_1 = \frac{(c - \tan\theta_0)\tan\theta_1}{1 - \tan\theta_0\tan\theta_1}.$$

Note that $\tan\theta_1 = \tan G(\theta_0) = \cot F(\theta_0)$. Since $\theta_0 < F(\theta_0)$, we have

$$0 < \tan\theta_0 \cot F(\theta_0) < 1.$$

The linearized eigenvalues of $\mathscr{R}$ at (x_0, θ_0) are given by

$$\frac{\partial Y}{\partial x}(x_0, \theta_0) = \cot F(\theta_0)\cot F(\theta_1), \quad \frac{dG^2}{d\theta}(\theta_0).$$

As $\cot F(\theta_0)\cot F(\theta_1) = \tan\theta_1 \cot F(\theta_1) < 1$, stability is determined by the stability of the angular dynamics. □

The theorems above tell us that the stability of rectangular or parallelogram-shaped cycles are determined by the angular dynamics and that their existence is dependent on c. Their stability does not depend upon c.

Example 5.15 (*Cubic function*) Theorems 5.13 and 5.14 generalize Theorems 5.2 and 5.4. We can observe rectangular and parallelogram-shaped cycles if F is a cubic function. Recall (5.8) in Example 5.9. Compare Fig. 5.15 with Fig. 5.10. We can observe a parallelogram-shaped cycle for $a = 0.24$. In this case, G has a periodic orbit $\{0.71558\ldots, 0.35277\ldots\}$. Note that θ_0 must be $0.71558\ldots$. In fact, if we take $\theta_0 = 0.35277\ldots$, then x_0 is negative.

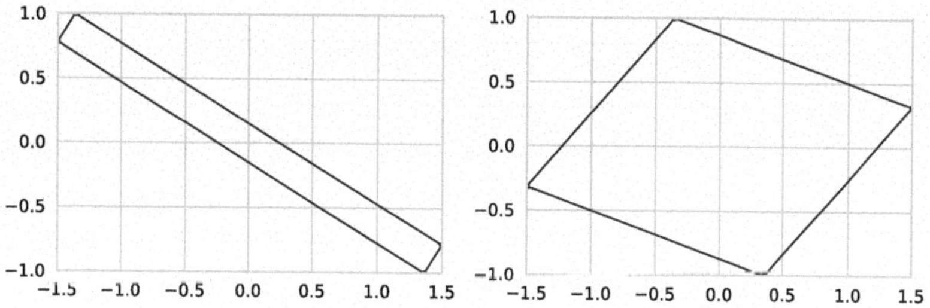

(a) $a = 0.2$ (b) $a = 0.24$

Fig. 5.15 Rectangle and parallelogram-shaped cycles for a cubic function

It is possible to discuss cycles of different types in a similar manner. Let us consider a figure-eight-shaped cycle for larger c. See Fig. 1.20d. Suppose that a particle hits the line segments AB, DC, CB, and BA in the following manner as shown in Fig. 5.16:

- The first reflection point divides AB into $x_0 : (c - x_0)$ ratio.
- The second point divides DC into $x_1 : (c - x_1)$ ratio.
- The third point divides CB into $x_2 : (1 - x_2)$ ratio.
- Finally, the fourth point divides BA into $x_3 : (c - x_3)$ ratio.

The angles of incidence are given by

$$\theta_1 = F(\theta_0), \quad \theta_2 = G(\theta_1), \quad \theta_3 = G(\theta_2),$$

and the numbers x_1, x_2, x_3 are given by

$$\begin{aligned} x_1 &= x_0 + \tan F(\theta_0), \\ x_2 &= (c - x_1) \cot F(\theta_1), \\ x_3 &= (1 - x_2) \cot F(\theta_2). \end{aligned}$$

If $\theta_3 = \theta_0$ and $x_3 = x_0$ hold, then the orbit is closed and forms a figure-eight-shape.

Theorem 5.16 *Suppose that F satisfies Assumptions 5.1–5.3. Let θ^* be a fixed point of $G^2 \circ F$ with $\theta^* \neq 0, \pi/2$. If c satisfies*

$$\tan F(\theta^*) + \cot F(G(F(\theta^*))) < c < \tan F(\theta) + \tan F^2(\theta),$$

then a figure-eight-shaped cycle exists.

Proof Let us consider a map $\mathscr{E}(x, \theta) = (Z(x, \theta), H(\theta))$ given by

$$\begin{aligned} Z(x, \theta) &= \left[1 - (c - x - \tan F(\theta)) \cot F^2(\theta)\right] \cot F(G(F(\theta))), \\ H(\theta) &= G^2(F(\theta)). \end{aligned}$$

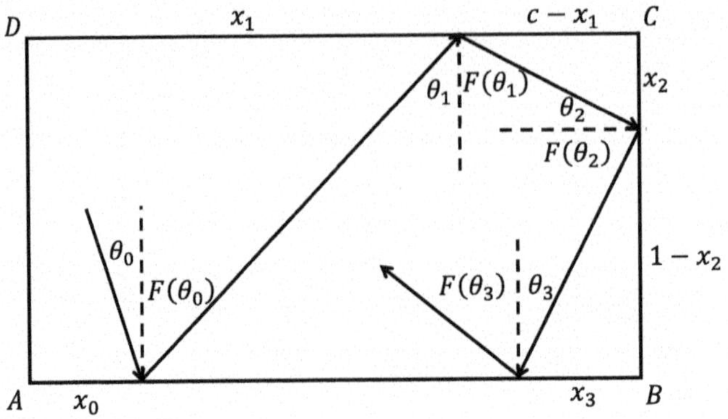

Fig. 5.16 Illustration of a figure-eight map

Since F and G^2 are monotonically increasing, H is also monotonically increasing. Therefore, the sequence $\theta_n = H^n(\theta_0)$ is monotone and cannot be periodic. Let $\tilde{H}(\theta) = H(\theta) - \theta$. We have $\tilde{H}(0) = 0$, $\tilde{H}(\pi/2) = 0$, and

$$\tilde{H}'(0) = (F'(0))^2 - 1, \quad \tilde{H}'\left(\frac{\pi}{2}\right) = F'(0) - 1.$$

Since we have $F'(0) > 1$, $\tilde{H}'(0)$ and $\tilde{H}'(\pi/2)$ are both positive. Therefore, there exists an $\varepsilon > 0$ such that $\tilde{H}(\varepsilon) > 0$ and $\tilde{H}(\pi/2 - \varepsilon) < 0$. By the intermediate value theorem, it turns out that H has a nontrivial fixed point. Let θ^* be this fixed point. Now, the figure-eight-shaped cycle exists if and only if $x_3 = x_0$ holds. The region of existence of the cycle is defined by

$$0 < x_0 < c, \quad 0 < x_1 < c, \quad 0 < x_2 < 1.$$

□

Strictly speaking, $0 < x_0 < c/2 < x_1 < c$ should be satisfied because the trajectory may have a self-intersection. The condition for avoiding such a self-intersection is left as an exercise.

5.4 Discrete-Time Model and Particle Model

In this section, we compare the particle model and the discrete model. We consider the particle model with $m_0 = m_1 = m_2 = a = 1$ and $\delta = 0.06$. The aspect ratio c and the domain size L are varied as controlling parameters.

Let us consider the angle relation for the particle model with $m_j = a = 1, \delta = 0.06$. Figures 5.17, 5.18, and 5.19 present square, rectangle, and eight-shaped cycles

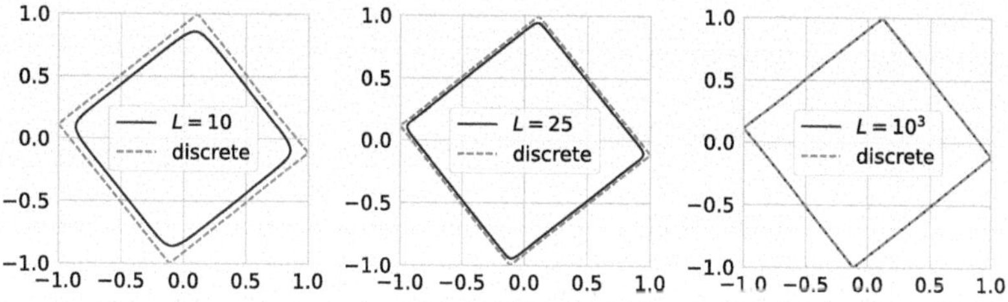

(a) $L = 10$ (b) $L = 25$ (c) $L = 10^3$

Fig. 5.17 Square-shaped cycle for particle model and discrete model. The horizontal and vertical axes indicate x/L and y/L, respectively

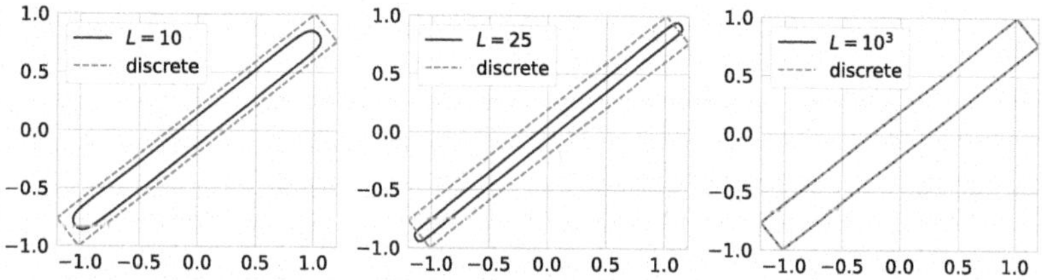

(a) $L = 10$ (b) $L = 25$ (c) $L = 10^3$

Fig. 5.18 Rectangle-shaped cycle for particle model and discrete model on rectangular domain with $c = 1.2$. The horizontal and vertical axes indicate x/L and y/L, respectively

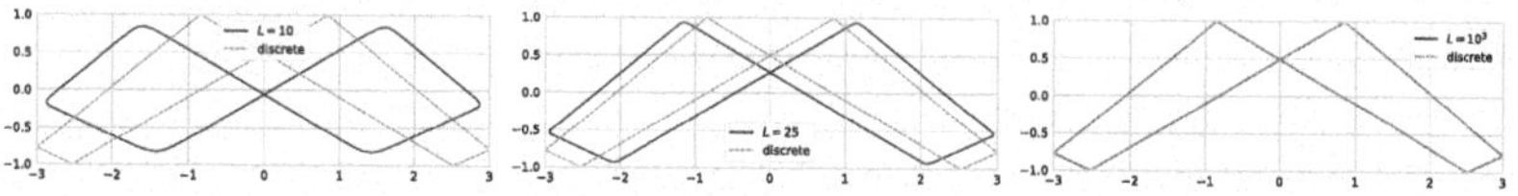

(a) $L = 10$ (b) $L = 25$ (c) $L = 10^3$

Fig. 5.19 Eight-shaped cycle for particle model and discrete model on rectangular domain with $c = 3.0$. The horizontal and vertical axes indicate x/L and y/L, respectively

for the particle model with $L = 10, 25, 10^3$ together with the corresponding cycle for the discrete model, respectively. We can see an evident gap between the orbits of the particle model with $L = 10$ and those of the discrete model. On the other hand, there is no visible gap for $L = 10^3$. This suggests that the discrete model provides us with a good approximation of the particle model for a large domain.

See Fig. 5.18b. The approach of the cycles of the particle model toward the discrete model is not monotone. Figure 5.20 shows a bifurcation diagram of the rectangular shaped cycle for the particle model with $c = 1.2$, where L is the controlling parameter. There are two saddle-node bifurcations at $L \approx 54.857$ and $L \approx 33.752$. Figure 5.21 shows three cycles at $L = 50$. The cycle on the upper branch has a rectangular shape similar to that displayed by the discrete model. On the other hand, the other two cycles are thinner than the rectangular cycle. Thus, the discrete-time model

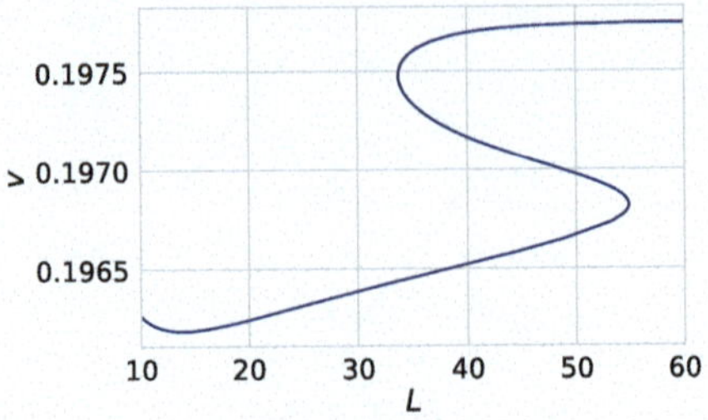

Fig. 5.20 Continuation of the cycle of particle model for $c = 1.2$ with respect to L

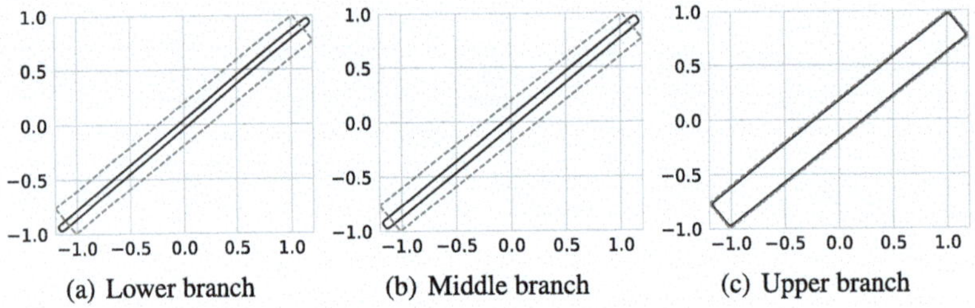

Fig. 5.21 Cycles for $L = 50$ (solid line) and the rectangle shaped cycle for the discrete model (dashed gray). The horizontal and vertical axes indicate x/L and y/L, respectively

does not necessarily approximate the particle model when the domain is mesoscopic, that is to say, medium sized.

Keeping this observation in mind, let us investigate the existence region of the rectangular-shaped cycles. Recall Theorem 5.2. For the angle relation shown in Fig. 5.3, we obtain $\theta^* \approx 0.67327$ and $c^* = \cot\theta^* \approx 1.25377$. In Fig. 4.12 we have seen a bifurcation diagram of a square-shaped cycle with respect to the aspect ratio c for the particle model with $m_j = a = 1$, $L = 10.0$, $\delta = 0.06$. There is a saddle-node bifurcation at $c \approx 1.630$. In addition, there are two saddle-node bifurcation points as shown in Fig. 5.20. We have applied methods of numerical continuation to follow the saddle-node bifurcation curves on the (c, L)-plane. The result is shown in Fig. 5.22. We can observe three asymptotes: the middle one is close to $c = c^*$, which is predicted by the discrete model. The other two asymptotes are not expected according to the discrete model. Figure 5.23 presents enlarged pictures of cycles for the particle model with $L = 500$ for $c = 1.21160, 1.24809$ and 1.76383. We can recognize Fig. 5.23c as rectangular. On the other hand, Fig. 5.23a, b are cigar-shaped with smooth rather than sharp corners. Their reflection points are too close to the corner of the domain. We cannot regard the part of the orbit near the reflection as a line segment.

Now let us observe the case where $c = 1.3$. See Fig. 5.24. The orbit of the particle model for $L = 10$ seems to be quasiperiodic or periodic with a long period. On the other hand, the attractor of the discrete time model is periodic. Indeed, the orbit returns to the initial position at the 35th reflection. For $L = 10^5$ we can see agreement

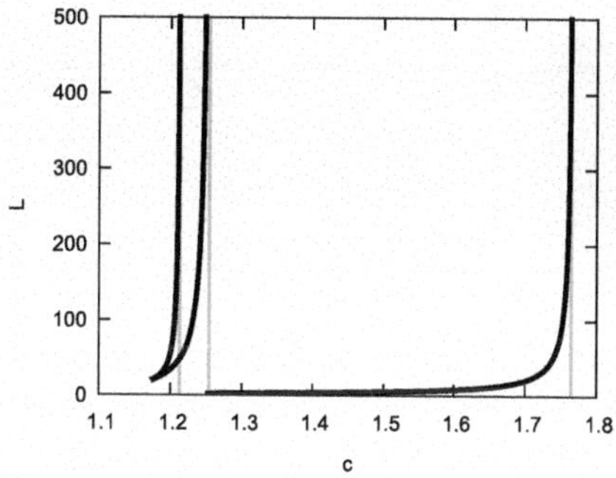

Fig. 5.22 Saddle-node bifurcation curves on the (c, L)-plane for the particle model with $m_j = a = 1, \delta = 0.06$

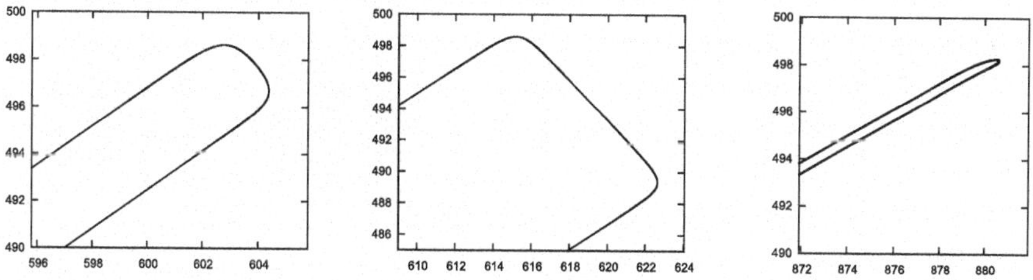

Fig. 5.23 Enlargements of cycle reflection characteristics for the particle model with $L = 500$. The horizontal and vertical axes indicate x and y, respectively

between the particle and discrete models. The convergence seems to be slower than in the cases of rectangular-shaped cycles. In fact, the cycle of the particle model for $L = 10^4$ is still far from that of the discrete model. In the case where $c = 1.3$, the attractor of the discrete model extends to the corners of the domain. Even though the trajectory of the particle model tends to the cycle of the discrete model, the particle can move in a different direction when it is close to the corners. The relation between the angles of incidence and reflection is independent of the site of reflection in the discrete model. On the other hand, in the particle model, the particle is affected by both edges near a corner, and for that reason, the motion of the particle differs from that predicted by the discrete model.

Figure 5.25 shows orbit diagrams for the particle model with $L = 10$ and $L = 10^3$ and the discrete model. We employed 2049 equally distributed points on the parameter interval $[1, 3]$. The vertical axis of Fig. 5.25 indicates $\arg(x + iy)$ for each reflection point (x, y). It would seem that the parameter regions of quasiperiodic or long period orbits are narrow for $L = 10^3$ and for the discrete model.

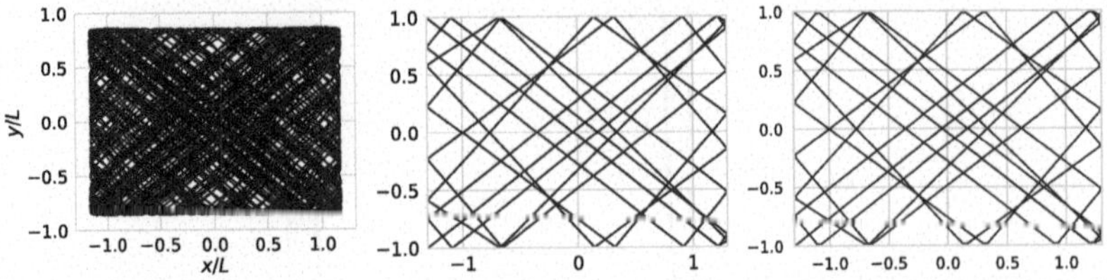

(a) $L = 10$ (b) $L = 10^5$ (c) discrete model

Fig. 5.24 Attractors for $c = 1.3$. The horizontal and vertical axes indicate x/L and y/L, respectively. Figure adapted from the author's previous work [49]

Fig. 5.25 Orbit diagrams for the particle model with $\delta = 0.06$, $m_j = a = 1$ and the discrete model. Figure adapted from the author's previous work [49]

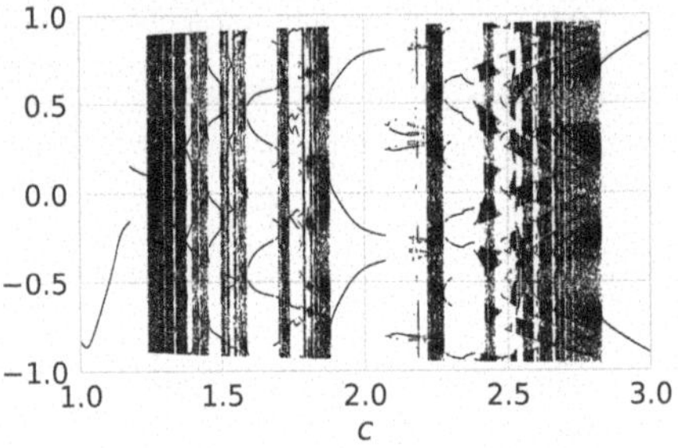

(a) $L = 10$

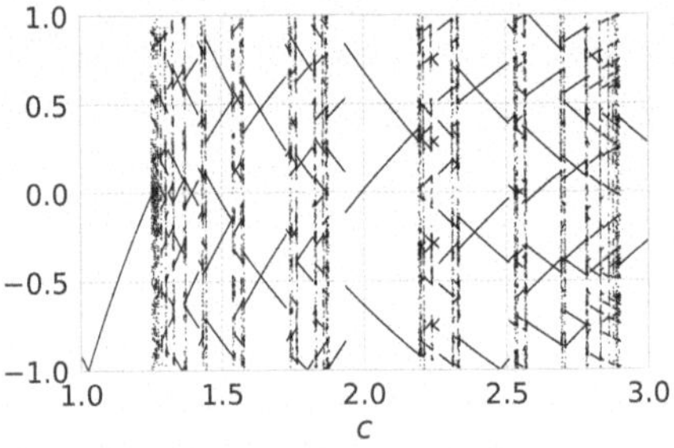

(b) $L = 10^3$

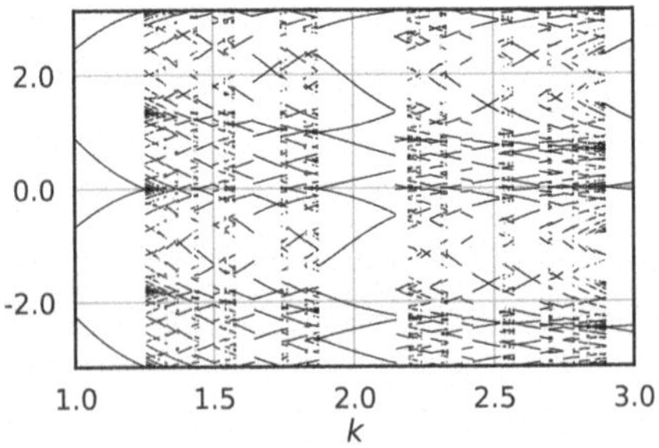

(c) discrete

In summary, the discrete model seems to provide a good approximation to the particle model for sufficiently large L. One can derive the existence and stability requirements using only elementary calculus. The numerical results suggest that the convergence rate depends on c. In particular, convergence depends on the period of the cycle for the discrete model. We have not observed a chaotic orbit for the discrete model. A modified discrete model has been proposed and numerically studied by Mimura et al. [49]. In that model, the angle of reflection depends not only upon the angle of incidence but also upon the site of reflection. It can produce a chaotic orbit in the sense that the largest Lyapunov exponent is positive. However, the mathematical analysis of the modified model is prohibitively difficult.

We have studied some cycles with simple shapes in the previous sections. It is possible to examine other cycles that have more complicated shapes and longer periods. The strategy described above may be summarized as follows. For a given cycle shape, we have defined a two-dimensional map and have found its fixed point. Since the angular dynamics are independent of the positional dynamics in the discrete model, we can treat the two separately. Once we find a fixed point of the angular dynamics, we obtain that of the positional dynamics via the solution of a linear equation. Although the formula may be quite complicated, one can analyze the existence and local stability of the cycle using elementary calculus, at least in principle.

In order to study the trajectory resulting from an arbitrary initial condition, we must define a two-dimensional dynamical system that deals with sites of reflection and velocity vectors. Suppose that the billiard table has a convex shape such as a square, rectangle, polygon, or circle. As it is homeomorphic to S^1, we can parameterize it by $\phi \in [0, 2\pi]$ modulo 2π. In addition, we can parameterize the velocity by $\theta \in [0, 2\pi]$ modulo 2π because speed is irrelevant here. Thus, the phase space is $S^1 \times S^1 \simeq \mathbb{T}^2$. It should be noted that there exist some singularities if the domain is a polygon. For example, suppose that $0, \pi/2, \pi, 3\pi/2$ are vertices of a square domain. What happens if a particle hits one of these vertices? In classical billiards, the particle reverses, following the same trajectory. This assumption is not, however, possible in our case because the dynamical system would become discontinuous. Such a dynamical system has yet to be studied.

5.5 Some Remarks

The discrete-time model was derived in a master's thesis written by Matsumoto [48]. In the thesis, sufficient conditions were provided for the existence and stability of square-shaped cycles on square domains and rectangular and figure-eight-shaped cycles on rectangular domains. Matsumoto also demonstrated that a square-shaped cycle can be unstable in the case where F is quadratic. Moreover, it had previously been thought that $F'(\pi/2) = 1$ is important for the stability of the square-shaped cycle. However, Matsumoto observed that a square-shaped cycle can lose stability even if $F'(\pi/2) = 1$ holds. For example, it has been numerically shown that a period-doubling bifurcation can occur in the case where F is quartic. In addition, Matsumoto investigated the case where F is piecewise linear.

Morihara has also studied the discrete-time model [52]. He has compared the trajectories of discrete-time models and the particle model in square domains that are not large, and has investigated the behavior of discrete-time models for some F. For $F(\theta) = \pi/2 - F_2(\pi/2 - \theta)$ with

$$F_2(\theta) = \begin{cases} \frac{1}{3}\theta & x \in [0, 3\pi/7), \\ 5\theta - 2\pi & x \in [3\pi/7, \pi/2], \end{cases}$$

typical orbits tend to a square-shaped cycle on a square domain. For $F(\theta) = \pi/2 - F_4(\pi/2 - \theta)$ with

$$F_4(\theta) = \begin{cases} \frac{1}{3}\theta & x \in [0, \pi/3), \\ 5\theta/2 - 3\pi/4 & x \in [\pi/3, \pi/2], \end{cases}$$

square-shaped cycles are unstable, and orbits tend to a figure-eight-shaped cycle on a square domain.

In classical billiards, the reflection rule is fixed, and numerous studies have been conducted on the subject of classical billiards on tables with various shapes. In this case, there is a certain degree of freedom in the selection of the reflection rule. As discussed above, there can be a rich structure depending on the reflection rule, even if the shape of the billiard table is fixed, although it is always tempting to think about billiard tables in shapes other than the standard rectangle.

If the domain is a disk, then the situation is rather simple. See Fig. 5.26. Suppose we have an orbit that reflects points A and B on the boundary of the domain. Because every line normal to the circle meets at the center, we have an isosceles triangle $\triangle OAB$. This implies that the next incident angle is equal to the current reflection angle. Hence, the angular dynamics are defined by iterations of F. Because F is monotonically increasing, $F^n(\theta)$ tends to $\pi/2$ as $n \to \infty$ for any $\theta \in (0, \pi/2)$. Hence, the orbit tends to be along the boundary of the domain.

Suppose the domain is a regular polygon. Let N be the number of edges of the domain. Now consider a periodic orbit that reflects from all the edges of the domain in turn. Suppose that the initial angle of incidence is θ_0. We then obtain the next angle of incidence θ_1 as

$$\theta_1 = \left(1 - \frac{2}{N}\right)\pi - F(\theta_0).$$

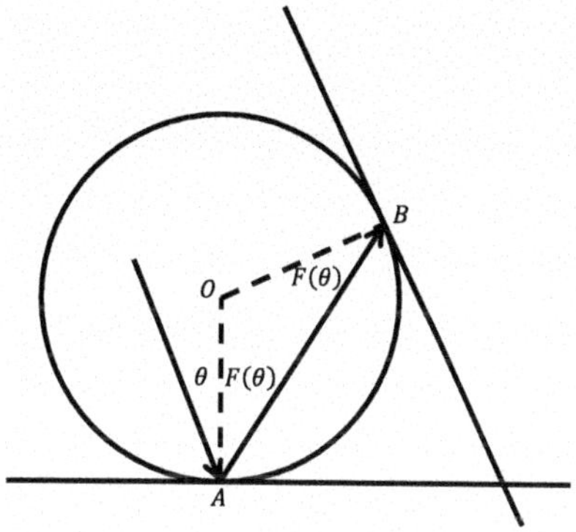

Fig. 5.26 Billiard motion in a disk. O is the center of the circle. The orbit reflects at A and B in turn. The two straight lines are tangent to the circle at A and B, respectively. The two dashed line segments are the normal lines at A and B

Note that we have $\theta_1 = \pi/2 - F(\theta_0)$ for $N = 4$. Let $G(\theta) = \left(1 - \frac{2}{N}\right)\pi - F(\theta)$. G must have a fixed point in $(0, \pi/2)$.

Remark 5.17 The angular dynamics for a circle domain are not derived from the limit $N \to \infty$. Indeed, $G(\theta) \to \pi - F(\theta)$ as $N \to \infty$. However, the angular dynamics in the circle domain are defined only by iterations of F.

Suppose the domain is an equilateral triangle. One can consider a periodic orbit that reflects off all the edges of the domain in turn.

$$x_{n+1} = (1 - x_n)\frac{\cos F(\theta_n)}{\sin\left(\frac{\pi}{6} + F(\theta_n)\right)}, \quad \theta_{n+1} = \frac{\pi}{3} - F(\theta_n).$$

There exists a fixed point for the angular dynamics. If θ^* is the fixed point, then it satisfies $0 < \theta^* < \pi/6$.

Remark 5.18 (*"Blackhole"*) Let us consider the reflections in a V-shaped region with an angle φ. Suppose that $0 < \varphi < \pi/2$ holds. Let θ_n be the angle of incidence at the nth collision. It is easy to see that θ_{n+1} is given by

$$\theta_{n+1} = F(\theta_n) - \varphi.$$

If there is an $m \in \mathbb{N}$ such that θ_m becomes negative, then the orbit escapes from the corner. For example, in the case of $F(\theta) = \theta$, the orbit goes outside after the at most $\lceil \pi/\varphi \rceil$th collision [78, Chap. 1], where $\lceil x \rceil$ is the smallest integer not less than x for $x \in \mathbb{R}$.

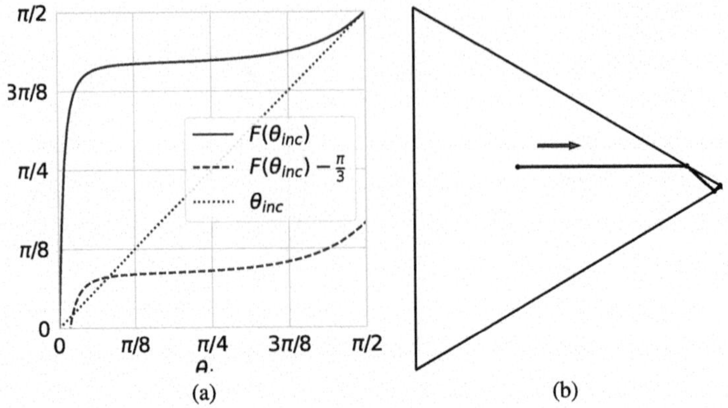

Fig. 5.27 Illustration of Remark 5.18. **a** The angle relation calculated from the particle model with $\delta = 0.6$, $a = 2$, $m_0 = m_1 = 1$. **b** An orbit in an equilateral triangle

Suppose that there exists θ^* satisfying $\theta^* = F(\theta^*) - \varphi$. If θ_n converges to θ^* satisfying $F'(\theta_n) > 0$, then the orbit cannot escape from the corner. The angle function calculated from the particle model with $\delta = 0.6$, $a = 2$, $m_0 = m_1 = 1$ satisfies this condition. See Fig. 5.27. In contrast, any orbit will escape from the corner in the particle model.

Chapter 6
Conclusion

Despite this book's title, we have only focused on camphor disk motion. There are other dissipative systems that involve billiard-like motion with nonspecular reflection. One of the classical examples is a droplet walker [6, 69]. Under certain conditions, a liquid droplet moves in a straight line and reflects from a tank wall at a nonspecular angle. Shirokoff has demonstrated by a mathematical model that the trajectory of the droplet approaches either non-periodic dense curves or a quasiperiodic orbit in a large domain [73]. Altmann et al. [3] have studied some optical systems in which the site of reflection shifts and the angle of reflection is greater than that of incidence. In [3], the authors have employed a discrete-time model to study a billiard problem in an annular domain, where reflections at the outer boundary are specular, and reflections at the inner boundary are non-specular. Certain microorganisms exhibit similar behavior. Kantsler et al. have demonstrated that *Chlamydomonas* approaching a flat wall moves away from the wall at a fixed angle, independent of the angle of incidence [33]. Similarly to Altmann et al.'s system, the site of reflection shifts when it reflects from a boundary wall. Spagnolie et al. [74] have studied a billiard problem in regular polygons using a discrete-time model which exhibits a fixed angle of reflection and shift of reflection sites. They have demonstrated stable cycles and chaotic orbits in regular polygons and trajectories in the presence of obstacles. Uspal et al. [80] have studied the motion of Janus particles, which are self-driven by interaction from hydrodynamic and chemical fields induced by their motion. They have shown that three types of motion near a boundary can occur depending on parameters: reflection, sliding and hovering. Some mathematical models of a cavity soliton formulated in terms of a nonlinear Schrödinger equation and ODEs also exhibit billiard-like motion with nonspecular reflection, but this has not been realized in laboratory experiments [67]. In addition, cores of spiral patterns in some reaction-diffusion systems also exhibit similar motions [44].

T. Miyaji et al., *A Billiard Problem in Nonlinear Dissipative Systems*,
Surveys and Tutorials in the Applied Mathematical Sciences 18,
https://doi.org/10.1007/978-981-95-7561-9_6

If such phenomena are modeled by dynamics of a localized structure described by some partial differential equations, a reduction theory similar to Chap. 3 can be applied to derive a particle model. A discrete-time model may be applicable if the domain is large enough. Thus, there are many opportunities to apply what has been discussed in the present book.

The MB model consists of two equations: an equation of motion of a rigid disk and a reaction-diffusion equation. The center manifold theory for a sufficiently slow traveling spot solution plays an essential role in the derivation of the particle model [9]. This theory is also applicable to planar reaction-diffusion systems [16]. The cavity soliton billiards [67] are similar to traveling spots in planar RDSs. Let us introduce their model without giving details. Prati et al. [67] proposed the following system of PDE and ODEs:

$$\begin{cases} \dfrac{\partial F}{\partial t} = \left[(1-\mathrm{i}\alpha)D + (1-\mathrm{i}\beta)d - 1 + \mathrm{i}\Delta\right]F, \\ \dot{D} = b_1\left[\mu - D(1+|F|^2) - BD^2\right], \\ \dot{d} = b_2\left[-\gamma - d(1+s|F|^2) - Bd^2\right], \end{cases}$$

where F, D, and d are function of time t and the planar coordinates (x, y). Here, $\mathrm{i}\Delta$ represents not a diffusion but a diffraction. Numerical study by Prati et al. [68] showed that this system has a spatially localized stationary solution called a cavity soliton (CS) in some parameter region and that the CS can undergo a drift bifurcation. Therefore, a moving CS is observed under certain conditions. In [67] they assumed that μ is a function $\mu = \mu(x, y)$ satisfying

$$\mu(x, y) = (\mu_{in} + \Delta\mu)e^{-(x^n+y^n)/w_2^n} - \Delta\mu e^{-(x^n+y^n)/w_1^n},$$

where they chose parameters so that $\mu(x, y)$ is above the threshold of the drift bifurcation in an almost square-shaped area and is below the threshold outside of it. By this function μ, a moving CS is confined in a finite domain. Prati et al. [67] have found that the moving CS tends to a square-shaped cycle. They have also studied the existence and stability of the square-shaped cycle based on observation of the relationship between angles of incidence and reflection. Although the cavity soliton billiards model is a kind of nonlinear Schrödinger equation, its stationary solution is given by a nonlinear Poisson equation in a similar manner to RDSs. Whether two CSs interact repulsively is of important. There are some numerical studies on interaction between two stationary CSs [4, 19, 71]. We expect that the framework of [16] is applicable, though there may be some technical difficulties in verifying existence and spectral properties of the CS from the view point of PDE theory.

The droplet walkers were introduced in [10, 69]. They investigated motions of an oil droplet bouncing on the surface of a vertically vibrated bath. The droplet moves by interacting with surface waves on the bath, and waves are produced by the bouncing droplet. Despite the fact that the droplet is of millimeter scale, it possesses a pilot wave, reminiscent of what was proposed by de Broglie in the field of quantum mechanics. Indeed, many aspects of particle-wave duality have been studied [66]. Bush et al. [8] have developed a mathematical model of the walkers and have demonstrated that the motion of a walker is described by equations of motion with speed-dependent mass:

$$\frac{d}{dt}\gamma_B(v)m\dot{\boldsymbol{v}} = D(v)\boldsymbol{v} + \boldsymbol{F},$$

where $v = |\boldsymbol{v}|$, $\gamma_B(v)$ is the speed-dependent mass, $\boldsymbol{F}$ is an external force, and $D(v) > 0$ if v is below a certain threshold or otherwise $D(v) < 0$. Pucci et al. [70] have studied reflections of walkers from boundary walls by experiments and numerical simulations. They have found that the angles of incidence and reflections are different and that the angle of reflection is typically in the range from 60° to 80°. In addition, they have suggested that $\boldsymbol{F}$ is perpendicular to the boundary wall. As the walkers consist of a droplet and surface waves, they are spatially extended structures rather than localized ones. Hence, the reduction theory in the present book is not straightforwardly applicable to the droplet walkers. However, the particle model is similar to the equation above.

Spot-like localized structures in one-dimensional reaction-diffusion systems are pulse solutions. Similarly, localized structures and their dynamics are observed in convection patterns in fluid dynamics [39, 40, 85] and cavity solitons in nonlinear optics [46]. It is believed that there are universal properties common to such spatial localized structures, which have been studied in various fields [1, 2, 45]. In this context, localized structures are called dissipative solitons. Some of the examples above would fall into the category of dissipative solitons. Alternatively, the reader may find universal properties in billiard problems from the point of view of dissipative systems.

Our study began with an experiment conducted by Mr. Kanda, who was a former student of the third author, which revealed the motions of a camphor disk moving on a water surface like a billiard ball. We have discussed three mathematical models throughout this book. Each model describes the same phenomenon but at different scales. The moving boundary (MB) model, which was discussed in Chap. 3, models the mechanism that drives camphor motion. By using this model, we have succeeded in simulating the billiard-like motion of a camphor disk. On the other hand, it is difficult to mathematically analyze the orbits of the disk of the MB model. To overcome this difficulty, we have assumed that the motion of the disk is so slow that the reduction to ordinary differential equations is meaningful. We have analyzed the particle model in Chap. 4. Although the particle model helps us to simulate and understand the motions of the disk on a rectangular domain, it is still difficult to study the existence and stability of limit cycles far from a bifurcation point. We have therefore imposed a further assumption to simplify the problem. That is, we have assumed that the domain (billiard table) is extremely large ($L \gg 1$). In such a situation, the motion is governed by a discrete-time model. For the study of limit cycles, the problem is essentially reduced to the existence and stability of a fixed point of a one-dimensional map. This is relatively straightforward to analyze, to the extent that one can deal with it by elementary calculus, as was shown in Chap. 5.

One may choose from the mathematical models developed here in accordance with the issues one wishes to look into more deeply. If one is interested in the phenomenological aspects of camphor motion, one can select the MB model even if it is hard to analyze. If one is interested in the analysis of limit cycles, one may prefer the discrete-time model even though this model is more abstract than the MB model. Someone specializing in the theory of ordinary differential equations may prefer the particle model. The particle model and the discrete-time model involve only a part of the phenomenon, simplifying the analysis. However, these two models have generality and applicability to other systems of self-propelling particles that continue moving in a straight line and exhibit repulsive interaction with the boundary. Generality is a benefit of abstraction. Thus, we have different models for different idealizations, and the capabilities required for analysis are also different for each model. These are part of the fun of mathematical modeling. In which model is the reader interested?

In sum, this book is a work of mathematical modeling and analysis, which has been inspired by experiments. We have found, through simulation and mathematical analysis, that camphor disk motions in a rectangular domain are quite different from those of classical mathematical billiards. In the context of nonlinear science, these are in the first instance just mathematical conjectures. In all honesty, it is still an open problem whether the reflections of an actual camphor disk are the same as those we have predicted. The reflections of actual camphor disks have not yet been studied experimentally to an extent which would allow reliable comparisons. In the last two decades, experiments on self-propelling particles have progressed apace. The authors will be pleased if future experiments verify or even disprove the conjectures that we have put forward.

References

1. Akhmediev, N., Ankiewicz, A. (eds.): Dissipative Soliton. In: Lecture Notes in Physics, vol. 661. Springer, Berlin, Heidelberg (2005)
2. Akhmediev, N., Ankiewicz, A. (eds.): Dissipative Solitons: From Optics to Biology and Medicine. In: Lecture Notes in Physics, vol. 751. Springer, Berlin, Heidelberg (2008)
3. Altmann, E.G., Magno, G.D., Hentschel, M.: Non-Hamiltonian dynamics in optical microcavities resulting from wave-inspired corrections to geometric optics. Europhys. Lett. **84**, 10008 (2008). https://doi.org/10.1209/0295-5075/84/10008
4. Anbardan, S.R., Rimoldi, C., Kheradmand, R., Tissoni, G., Prati, F.: Exponentially decaying interaction potential of cavity solitons. Phys. Rev. E **97**, 032208 (2018). https://doi.org/10.1103/PhysRevE.97.032208
5. Arnol'd, V., Afraimovich, V., Il'yashenko, Y., Shil'nikov, L.: Bifurcation Theory. Springer, Berlin, Heidelberg (1994)
6. Borghesi, C., Moukhtar, J., Labousse, M., Eddi, A., Fort, E., Couder, Y.: Interaction of two walkers: wave-mediated energy and force. Phys. Rev. E **90**, 063017 (2014). https://doi.org/10.1103/PhysRevE.90.063017
7. Brezis, H.: Functional Analysis, Sobolev Spaces and Partial Differential Equations. Springer, New York (2010)
8. Bush, J.W.M., Oza, A.U., Moláček, J.: The wave-induced added mass of walking droplets. J. Fluid Mech. **755**, R7 (2014). https://doi.org/10.1017/jfm.2014.459
9. Chen, X., Ei, S.-I., Mimura, M.: Self-motion of camphor discs, model and analysis. Netw. Heterog. Media **4**(1) 1–18 (2009). https://doi.org/10.3934/nhm.2009.4.1
10. Couder, Y., Protière, S., Fort, E., Boudaoud, A.: Walking and orbiting droplets. Nature **437**, 208 (2005). https://doi.org/10.1038/437208a
11. Das, M., Schmidt, C.F., Murrell, M.: Introduction to active matter. Soft Matter **16**, 7185–7190 (2020). https://doi.org/10.1039/D0SM90137G
12. Doedel, E.J., Oldeman, B.E.: AUTO07-P: continuation and bifurcation software for ordinary differential equations. http://indy.cs.concordia.ca/auto/
13. Ei, S.-I.: The motion of weakly interacting pulses in reaction-diffusion systems. J. Dyn. Differ. Equ. **14**(1), 85–137 (2002). https://doi.org/10.1023/A:1012980128575
14. Ei, S.-I., Kitahata, H., Koyano, Y., Nagayama, M.: Interaction of non-radially symmetric camphor particles. Phys. D **366**, 10–26 (2018). https://doi.org/10.1016/j.physd.2017.11.004
15. Ei, S.-I., Mimura, M., Miyaji, T.: Reflection of a self-propelling rigid disk from a boundary. Discrete Contin. Dyn. Syst. Ser. S **14**(3), 803–817 (2021). https://doi.org/10.3934/dcdss.2020229

T. Miyaji et al., *A Billiard Problem in Nonlinear Dissipative Systems*, Surveys and Tutorials in the Applied Mathematical Sciences 18, https://doi.org/10.1007/978-981-95-7561-9

16. Ei, S.-I., Mimura, M., Nagayama, M.: Interacting spots in reaction diffusion systems. Discrete Contin. Dyn. Syst. **14**(1), 31–62 (2006). https://doi.org/10.3934/dcds.2006.14.31
17. Ei, S.-I., Ohta, T.: Equation of motion for interacting pulses. Phys. Rev. E **50**, 4672 (1994). https://doi.org/10.1103/PhysRevE.50.4672
18. Ei, S.-I., Wei, J.: Dynamics of metastable localized patterns and its application to the interaction of spike solutions for the Gierer–Meinhardt systems in two spatial dimension. Jpn. J. Ind. Appl. Math. **19**(2), 181–226 (2002). https://doi.org/10.1007/BF03167453
19. Eslami, M., Khiavi, N.H., Kheradmand, R., Prati, F.: Localized optical micromotors based on twin laser cavity solitons. Phys. Rev. A **98**, 043807 (2018). https://doi.org/10.1103/PhysRevA.98.043807
20. Frigo, M., Johnson, S.: The design and implementation of FFTW3. Proc. IEEE **93**(2), 216–231 (2005). https://doi.org/10.1109/JPROC.2004.840301
21. Furuya, E.: Numerical computation of a phase field model describing motions of a camphor disk (in Japanese). Master's thesis, Hiroshima University (2001)
22. Golubitsky, M., Stewart, I.: The Symmetry Perspective. Birkhäuser, Basel (2002). https://doi.org/10.1007/978-3-0348-8167-8
23. Golubitsky, M., Stewart, I., Schaeffer, D.: Singularities and Groups in Bifurcation Theory, vol. II. Springer, New York (1988). https://doi.org/10.1007/978-1-4612-4574-2
24. Guckenheimer, J., Holmes, P.: Nonlinear Oscillations, Dynamical Systems, and Bifurcations of Vector Fields. Springer, New York (1983). https://doi.org/10.1007/978-1-4612-1140-2
25. Haragus, M., Iooss, G.: Local Bifurcations, Center Manifolds, and Normal Forms in Infinite Dimensional Dynamical Systems. Springer, London (2011). https://doi.org/10.1007/978-0-85729-112-7
26. Hayashima, Y., Nagayama, M., Nakata, S.: A camphor grain oscillates while breaking symmetry. J. Phys. Chem. B **105**(22), 5353–5357 (2001). https://doi.org/10.1021/jp004505n
27. Henry, D.: Geometric Theory of Semilinear Parabolic Equations. In: Lecture Notes in Mathematics, vol. 840. Springer, Berlin, Heidelberg (1981). https://doi.org/10.1007/BFb0089647
28. Hirsch, M.W., Smale, S., Devaney, R.L.: Differential Equations, Dynamical Systems, and an Introduction to Chaos, 3rd edn. Academic Press, Waltham (2013). https://doi.org/10.1016/C2009-0-61160-0
29. Hochbruck, M., Ostermann, A.: Exponential integrators. Acta Numer. **19**, 209–286 (2010). https://doi.org/10.1017/S0962492910000048
30. Ikeda, K., Ei, S.-I.: Center manifold theory for the motions of camphor boats with delta function. J. Dyn. Differ. Equ. **33**, 621–657 (2021). https://doi.org/10.1007/s10884-020-09824-9
31. Ikura, Y.S., Tenno, R., Kitahata, H., Suematsu, N.J., Nakata, S.: Suppression and regeneration of camphor-driven Marangoni flow with the addition of sodium dodecyl sulfate. J. Phys. Chem. B **116**(3), 992–996 (2012). https://doi.org/10.1021/jp210990a
32. Kanda, Y.: Experiments and numerical analyses for motions of a camphor disk (in Japanese). Bachelor thesis, Hiroshima University (2002)
33. Kantsler, V., Dunkel, J., Polin, M., Goldstein, R.E.: Ciliary contact interactions dominate surface scattering of swimming eukaryotes. Proc. Natl. Acad. Sci. USA **110**(4), 1187–1192 (2013). https://doi.org/10.1073/pnas.1210548110
34. Kassam, A.K., Trefethen, L.N.: Fourth-order time-stepping for stiff PDEs. SIAM J. Sci. Comput. **26**(4), 1214–1233 (2005). https://doi.org/10.1137/S1064827502410633
35. Kawasaki, K., Ohta, T.: Kink dynamics in one-dimensional nonlinear systems. Phys. A **116**(3), 573–593 (1982). https://doi.org/10.1016/0378-4371(82)90178-9
36. Kitahata, H., Koyano, Y., Iida, K., Nagayama, M.: Mathematical model and analyses on spontaneous motion of camphor particle, chap. 2. In: Nakata, S., Pimienta, V., Lagzi, I., Kitahata, H., Suematsu, N.J. (eds.) Self-organized Motion: Physicochemical Design Based on Nonlinear Dynamics, pp. 31–62. Royal Society of Chemistry, London (2019). https://doi.org/10.1039/9781788013499-00031
37. Kitahata, H., Yoshikawa, K.: Chemo-mechanical energy transduction through interfacial instability. Phys. D **205**(1), 283–291 (2005). https://doi.org/10.1016/j.physd.2004.12.012

38. Knobloch, E.: Spatial localization in dissipative systems. Annu. Rev. Condens. Matter Phys. **6**(1) 325–359 (2015). https://doi.org/10.1146/annurev-conmatphys-031214-014514
39. Kolodner, P.: Collisions between pulses of traveling-wave convection. Phys. Rev. A **44**, 6466–6479 (1991). https://doi.org/10.1103/PhysRevA.44.6466
40. Kolodner, P., Bensimon, D., Surko, C.M.: Traveling-wave convection in an annulus. Phys. Rev. Lett. **60** 1723–1726 (1988). https://doi.org/10.1103/PhysRevLett.60.1723
41. Koyano, Y., Yoshinaga, N., Kitahata, H.: General criteria for determining rotation or oscillation in a two-dimensional axisymmetric system. J. Chem. Phys. **143**(1), 014117 (2015). https://doi.org/10.1063/1.4923421
42. Krogstad, S.: Generalized integrating factor methods for stiff PDEs. J. Comput. Phys. **203**(1), 72–88 (2005). https://doi.org/10.1016/j.jcp.2004.08.006
43. Kuznetsov, Y.A.: Elements of Applied Bifurcation Theory, 3rd edn. Springer, New York (2004). https://doi.org/10.1007/978-1-4757-3978-7
44. Langham, J., Barkley, D.: Non-specular reflections in a macroscopic system with wave-particle duality: spiral waves in bounded media. Chaos **23**(1), 013134 (2013). https://doi.org/10.1063/1.4793783
45. Liehr, A.W.: Dissipative Solitons in Reaction Diffusion Systems, Mechanisms, Dynamics, Interaction. Springer, Berlin Heidelberg (2013). https://doi.org/10.1007/978-3-642-31251-9
46. Lugiato, L.A., Prati, F., Gorodetsky, M.L., Kippenberg, T.J.: From the Lugiato–Lefever equation to microresonator-based soliton Kerr frequency combs. Philos. Trans. R. Soc. A **376**, 20180113 (2018). https://doi.org/10.1098/rsta.2018.0113
47. Matsumoto, M.: Analysis of the particle model describing motions of a camphor (in Japanese). Master's thesis, Hiroshima University (2002)
48. Matsumoto, T.: A billiard problem under nonlinear and nonequilibrium conditions (in Japanese). Master's thesis, Hiroshima University (2003)
49. Mimura, M., Miyaji, T., Ohnishi, I.: A billiard problem in nonlinear and nonequilibrium systems. Hiroshima Math. J. **37**(3), 343–384 (2007). https://doi.org/10.32917/hmj/1200529808
50. Miyaji, T.: Arnold tongues in a billiard problem in nonlinear and nonequilibrium systems. Phys. D **340**, 14–25 (2017). https://doi.org/10.1016/j.physd.2016.09.003
51. Miyaji, T., Sinclair, R.: Asymptotic reflection of a self-propelled particle from a boundary wall. Jpn. J. Ind. Appl. Math. **41**, 269–295 (2024). https://doi.org/10.1007/s13160-023-00602-w
52. Morihara, T.: A nonequilibrium billiard problem: simulations and analyses (in Japanese). Master's thesis, Hiroshima University (2004)
53. Nagayama, M., Nakata, S., Doi, Y., Hayashima, Y.: A theoretical and experimental study on the unidirectional motion of a camphor disk. Phys. D **194**(3), 151–165 (2004). https://doi.org/10.1016/j.physd.2004.02.003
54. Nagayama, M., Monobe, H., Sakakibara, K., Nakamura, K.-I., Kobayashi, Y., Kitahata, H.: On the reaction-diffusion type modelling of the self-propelled object motion. Sci. Rep. **13**, 12633 (2023). https://doi.org/10.1038/s41598-023-39395-w
55. Nakata, S., Hayashima, Y.: Spontaneous dancing of a camphor scraping. J. Chem. Soc. Faraday Trans. **94**, 3655–3658 (1998). https://doi.org/10.1039/A806281A
56. Nakata, S., Hayashima, Y., Komoto, H.: Spontaneous switching of camphor motion between two chambers. Phys. Chem. Chem. Phys. **2**, 2395–2399 (2000). https://doi.org/10.1039/B000648N
57. Nakata, S., Iguchi, Y., Ose, S., Kuboyama, M., Ishii, T., Yoshikawa, K.: Self-rotation of a camphor scraping on water: new insight into the old problem. Langmuir **13**(16), 4454–4458 (1997). https://doi.org/10.1021/la970196p
58. Nakata, S., Nagayama, M.: Theoretical and experimental design of self-propelled objects based on nonlinearity, chap. 1. In: Nakata, S., Pimienta, V., Lagzi, I., Kitahata, H., Suematsu, N.J. (eds.) Self-organized Motion: Physicochemical Design Based on Nonlinear Dynamics, pp. 1–30. Royal Society of Chemistry, London (2019). https://doi.org/10.1039/9781788013499-00001
59. Nakata, S., Pimienta, V., Lagzi, I., Kitahata, H., Suematsu, N.J. (eds.): Self-organized Motion. Theoretical and Computational Chemistry Series. The Royal Society of Chemistry, London (2019). https://doi.org/10.1039/9781788013499

60. National Center for Biotechnology Information: PubChem compound summary for CID 2537, camphor. https://pubchem.ncbi.nlm.nih.gov/compound/Camphor (2025). Accessed 17 Mar 2025
61. National Center for Biotechnology Information: PubChem compound summary for CID 64685, borneol. https://pubchem.ncbi.nlm.nih.gov/compound/1_7_7-trimethylbicyclo_2.2.1_heptan-2-ol (2025). Accessed 17 Mar 2025
62. Nishiura, Y.: Far-from-Equilibrium Dynamics. Translations of Mathematical Monographs, vol. 209. American Mathematical Society, Providence (2002). https://doi.org/10.1090/mmono/209
63. Nishiura, Y., Teramoto, T., Ueda, K.-I.: Scattering of traveling spots in dissipative systems. Chaos **15**(4) (2005). https://doi.org/10.1063/1.2087127
64. Nishiura, Y., Teramoto, T.: Refraction, reflection and splitting: with thanks to all the people who allowed us to continue to make a research (far-from-equilibrium dynamics). RIMS Kokyuroku Bessatsu **B31**, 167–194 (2012). http://hdl.handle.net/2433/196226
65. Nishiura, Y., Teramoto, T., Yuan, X.: Heterogeneity-induced spot dynamics for a three-component reaction-diffusion system. Commun. Pure Appl. Anal. **11**(1), 307–338 (2012). https://doi.org/10.3934/cpaa.2012.11.307
66. Perrard, S., Labousse, M., Miskin, M., Fort, E., Couder, Y.: Self-organization into quantized eigenstates of a classical wave-driven particle. Nat. Commun. **5**, 3219 (2014). https://doi.org/10.1038/ncomms4219
67. Prati, F., Lugiato, L.A., Tissoni, G., Brambilla, M.: Cavity soliton billiards. Phys. Rev. A **84**, 053852 (2011). https://doi.org/10.1103/PhysRevA.84.053852
68. Prati, F., Tissoni, G., Lugiato, L.A., Aghdami, K., Brambilla, M.: Spontaneously moving solitons in a cavity soliton laser with circular section. Eur. Phys. J. D **59**, 73–79 (2010). https://doi.org/10.1140/epjd/e2010-00123-7
69. Protière, S., Boudaoud, A., Couder, Y.: Particle-wave association on a fluid interface. J. Fluid Mech. **554**, 85–108 (2006). https://doi.org/10.1017/S0022112006009190
70. Pucci, G., Sáenz, P.J., Faria, L.M., Bush, J.W.M.: Non-specular reflection of walking droplets. J. Fluid Mech. **804**, R3 (2016). https://doi.org/10.1017/jfm.2016.537
71. Rahmani Anbardan, S., Rimoldi, C., Kheradmand, R., Tissoni, G., Prati, F.: Interaction of cavity solitons on an unstable background. Phys. Rev. E **101**, 042210 (2020). https://doi.org/10.1103/PhysRevE.101.042210
72. Rozikov, U.A.: An Introduction to Mathematical Billiards. World Scientific, New Jersey (2018). https://doi.org/10.1142/11162
73. Shirokoff, D.: Bouncing droplets on a billiard table. Chaos **23**, 013115 (2013). https://doi.org/10.1063/1.4790840
74. Spagnolie, S.E., Wahl, C., Lukasik, J., Thiffeault, J.L.: Microorganism billiards. Phys. D **341**, 33–44 (2017). https://doi.org/10.1016/j.physd.2016.09.010
75. Strutt, R.J.: IV. Measurements of the amount of oil necessary IU order to check the motions of camphor upon water. Proc. R. Soc. Lond. **47**, 364–367 (1890). https://doi.org/10.1098/rspl.1889.0099
76. Suematsu, N.J., Nakata, S.: Evolution of self-propelled objects: from the viewpoint of nonlinear science. Chem. Eur. J. **24**(24), 6308–6324 (2018). https://doi.org/10.1002/chem.201705171
77. Swift, J.W.: Hopf bifurcation with the symmetry of the square. Nonlinearity **1**(2), 333–377 (1988). https://doi.org/10.1088/0951-7715/1/2/003
78. Tabachnikov, S.: Geometry and Billiards. American Mathematical Society, Providence (2005)
79. Tomlinson, C., Miller, W.A.: II. On the motions of camphor on the surface of water. Proc. R. Soc. Lond. **11**, 575–577 (1862). https://doi.org/10.1098/rspl.1860.0124
80. Uspal, W.E., Popescu, M.N., Dietrich, S., Tasinkevych, M.: Self-propulsion of a catalytically active particle near a planar wall: from reflection to sliding and hovering. Soft Matter **11**, 434–438 (2015). https://doi.org/10.1039/C4SM02317J
81. Vanag, V.K., Epstein, I.R.: Localized patterns in reaction-diffusion systems. Chaos **17**, 037110 (2007). https://doi.org/10.1063/1.2752494

82. Vanderbauwhede, A., Iooss, G.: Center manifold theory in infinite dimensions. In: Jones, C.K.R.T., Kirchgraber, U., Walther, H.O. (eds.) Dynamics Reported Expositions in Dynamical Systems New Series: Volume 1, pp. 125–163. Springer, Berlin, Heidelberg (1992). https://doi.org/10.1007/978-3-642-61243-5_4
83. Vicsek, T., Czirók, A., Ben-Jacob, E., Cohen, I., Shochet, O.: Novel type of phase transition in a system of self-driven particles. Phys. Rev. Lett. **75**, 1226–1229 (1995). https://doi.org/10.1103/PhysRevLett.75.1226
84. Vicsek, T., Zafeiris, A.: Collective motion. Phys. Rep. **517**(3), 71–140 (2012). https://doi.org/10.1016/j.physrep.2012.03.004
85. Watanabe, T., Iima, M., Nishiura, Y.: Spontaneous formation of travelling localized structures and their asymptotic behaviour in binary fluid convection. Fluid Mech. **712**, 219–243 (2012). https://doi.org/10.1017/jfm.2012.413

Index

T. Miyaji et al., *A Billiard Problem in Nonlinear Dissipative Systems*, Surveys and Tutorials in the Applied Mathematical Sciences 18,
https://doi.org/10.1007/978-981-95-7561-9

The manufacturer's authorised representative in the EU is Springer Nature Customer Service Centre GmbH, Europaplatz 3, 69115 Heidelberg, Germany. If you have any concerns regarding our products, please contact ProductSafety@springernature.com

Printed and bound by CPI Group (UK) Ltd, Croydon, CR0 4YY
09/10/2026
02230869-0001